5-9-74
Through Dr. Kennett
Dir Res. Tate & Lyle.

SWEETNESS AND SWEETENERS

An industry–university co-operation Symposium organised under the auspices of the National College of Food Technology, University of Reading, on 20th and 21st April, 1971

THE SYMPOSIUM COMMITTEE

GORDON G. BIRCH, B.Sc., Ph.D., F.R.I.C., M.R.S.H.
Lecturer at National College of Food Technology,
Reading University, Weybridge, Surrey.

LESLIE F. GREEN, B.Sc., A.R.I.C.
Glucose Projects Manager, Beecham Products,
Great West Road, Brentford, Middlesex.

C. BARRIE COULSON, M.Sc., Ph.D., Dr. Nat. Sc., F.R.I.C.
Reader at National College of Food Technology,
Reading University, Weybridge, Surrey.

H. VIZARD ROBINSON, A.C.I.S.
Secretary at National College of Food Technology,
Reading University, Weybridge, Surrey.

SWEETNESS AND SWEETENERS

Edited by

G. G. BIRCH, L. F. GREEN and C. B. COULSON

APPLIED SCIENCE PUBLISHERS LTD
LONDON

RIPPLE ROAD, BARKING, ESSEX, ENGLAND

ISBN 0 85334 503 1

LIBRARY OF CONGRESS CATALOG CARD NUMBER 78–168917
WITH 26 TABLES AND 57 ILLUSTRATIONS

Printed in Great Britain by Galliard Limited, Great Yarmouth, England

Main Participants in Symposium

L. M. BEIDLER
Department of Biological Science, The Florida State University, Tallahassee 32306, *Florida, USA.*

G. G. BIRCH and C. K. LEE
National College of Food Technology, University of Reading, St George's Avenue, Weybridge, Surrey.

A. J. COLLINGS
Environmental Safety Division, Unilever Research Laboratory, Colworth House, Sharnbrook, Bedfordshire.

P. S. ELIAS
*Department of Health and Social Security, Alexander Fleming House, London, SE*1.

L. F. GREEN
Research and Development Department, Beecham Products, Great West Road, Brentford, Middlesex.

SIR EDMUND HIRST
Department of Chemistry, University of Edinburgh.

R. M. HOROWITZ and BRUNO GENTILI
United States Department of Agriculture, Agriculture Research Service, Fruit and Vegetable Chemistry Laboratory, Pasadena, California 91106, *USA.*

G. E. INGLETT
Northern Utilisation Research and Development Division, Agricultural Research Service, US Department of Agriculture, Peoria, Illinois 61604, *USA.*

K. SELBY and J. TAGGART
The Lord Rank Research Centre, Lincoln Road, High Wycombe, Buckinghamshire.

R. S. SHALLENBERGER
New York State Agricultural Experimental Station, Cornell University, Geneva, New York 14456, *USA.*

H. W. SPENCER
Lyons Central Laboratories, 149 *Hammersmith Road, London, W.*14.

R. H. J. WATSON
*Department of Nutrition, Queen Elizabeth College, University of London, Campden Hill, London, W*8.

Contents

Opening Remarks

SIR EDMUND HIRST

The importance of sweetness as a factor in the palatability of foodstuffs has long been recognised and many attempts have been made to gain an understanding of the physical and chemical basis of the sensation. The problems involved have proved to be immensely complicated and many of their aspects are as yet far from being resolved. In the first place it has been found that in addition to certain sugars some 50 chemical substances covering a very wide variety of molecular structure exhibit this property in intensities, with sucrose taken as unity, ranging from very slight to 4000 or more. The sensitivity and selectivity of physiological response to taste and aroma are equally extraordinary and there is evidence from studies of insect behaviour that full response may result from the impact of one or two molecules of the specific chemical on the receptor organs. In man one need only refer to the amazing powers of discrimination possessed by some members of taste-panels, including experts in wine tasting. At the same time, however, these powers are intensely personal and at the other end of the scale there are individuals who fail completely to identify the quality of sweetness. Nevertheless, despite these difficulties, many attempts have been made during the past few years to correlate the attribute of sweetness with molecular structure. In this connection the work of Professor Shallenberger and his colleagues has been specially noteworthy and we look forward to hearing from him during the course of this symposium his views on the progress of these investigations. Other aspects of these problems will be discussed by Dr Beidler and by Dr Birch and we are to hear from Dr Inglett about the quite extraordinary intensity of sweetness possessed by certain natural products.

The simplest substances which display the property of sweetness possess molecular structures in which there are hydroxyl groups on two contiguous carbon atoms. In Professor Shallenberger's view substances of this type, including those which possess a hydroxyl group contiguous to a group such as carbonyl which can assume an acidic function, may exhibit sweetness provided the geometrical arrangement of the atoms is such that the necessary intimate association of the substances with the receptor organs becomes possible. How important these geometrical considerations are may be illustrated by the analogous case of enzyme activity where a rate of reaction may be enhanced by several powers of ten if, and only if, there is an exact fit between the molecular structures of the enzyme and the reactants. It is not as yet by any means certain that our present ideas of bonding are sufficient to provide an explanation of the phenomena.

Similar considerations may well apply in the case of bonding, whether by hydrogen bonding or otherwise, between the sweet substance and the taste buds, and it is of interest to think for a moment of the application of these views to the behaviour of simple sugars. As pointed out by Professor Shallenberger these molecules are much less rigid in structure than might be expected from their formal representations in textbooks. The β-form of D-glucose, for instance, tends to assume as its most stable form a structure in which the six carbon atoms occupy a chair-like ring, with these atoms and the six oxygen atoms present in only two planes—a very compact and stable conformation. Changes occur, however, immediately β-D-glucose comes into contact with an aqueous medium and some of the β-D-glucose is transformed into the α-variety, with alteration of distances between the hydrogens of the hydroxyl groups and the differing pattern of energy distribution between the numerous hydroxyl groups. In some instances, such as D-fructose, there may be in addition a ring change from the six-membered pyranose to the five-membered furanose type and, if circumstances demand it, a boat-shaped six membered ring may be formed. Other ring forms of intermediate shape may also occur and it is possible to have present some of the open chain form, the flexibility of which permits it to assume a wide variety of conformations. All these transformations require very little in the way of activation energy to enable them to take place, and they vary little between themselves in energy content. Conformational changes of these kinds can take place without alteration of bond angles or of bond lengths of any of the co-valent bonds in the sugar molecule. They do, however, provide a wide range of opportunities for different types of both intramolecular and intermolecular hydrogen bonding between the sugar molecules themselves and between sugar molecules and materials such as proteins. In view of the energy relationships such changes may readily take place at the call of a non-sugar molecule in order to bring about hydrogen or other types of bonding. It must be remembered also that some alteration in co-valent bond lengths or bond angles may accompany the other conformational changes, although here the energy requirements would be expected to be considerably greater.

A further extremely important point is that the changes we have in mind resulting in the close association of the sugar molecule with the receptor molecule, as envisaged by Professor Shallenberger, take place in an aqueous medium, which may, however, have properties very different from those of the ordinary dilute aqueous solutions of laboratory experiments. For here the water molecules are themselves closely associated through hydrogen bonding with the macromolecules of the polymer structure of the receptor. This may well result in an organised system of polymer and surrounding water molecules possessing solvent properties markedly different from those of water in a dilute aqueous solution. To add to the uncertainties there is the further fact that some sugars are unstable in the presence of proteins or aminoacids and may undergo extensive degradation with formation of numerous reaction products. Such reactions are normally slow and since the perception of sweetness appears to be almost instantaneous they are not likely to be of great importance in problems of the kind now under discussion. It would seem, nevertheless, that the number of factors which must necessarily be taken into account in an attempt to correlate sugar structure with potential sweetness is extremely large. It may well be the case that progress towards such a correlation may be found to be less difficult with more rigid molecules, for instance saccharin, which do not possess the extraordinary flexibility of the carbohydrate molecules, and Professor Shallenberger has already put forward some suggestions in this direction.

So far we have been considering only problems related to the physico-chemical constants between molecules of the sweet substance and those of the receptor in the taste buds. But this is really only the first stage in the enquiries to be dealt with in this symposium. Very little is known about the nature of the receptor units and still less concerning the nature of the interaction at this surface which results in the transmission of an electric impulse along the appropriate nerve. Then come problems of psychology as to how these impulses are decoded and interpreted as sensations of sweetness. In the second paper of the series Dr Watson is to speak to us on some of the problems involved in the psychology of sweetness.

Proceeding now to the use of sweeteners in foodstuffs, problems of many kinds emerge, most of them of prime importance. To take just one example, the organic chemist has produced substances with many times the sweetness of sucrose, fructose or other natural sugars, and these can be made at a cost competitive with that of sucrose. In view of the evidence which is now accumulating to the effect that the vast consumption of sucrose in the western world is in many instances a hazard to health, is it practicable to retain the desired sweetness of foodstuffs by the use of synthetic sweetness? The case of cyclamates which recently caused so great a disturbance will serve as illustration. Here was a substance which had undergone toxicological tests and had been in use for a considerable time, but on the basis of still more rigorous testing was judged not to be free from danger and was banned. The whole question of the testing of artificial sweeteners for toxicity is of extreme urgency. Complications due to the impossibility of testing on man and the difficulty of being able with any degree of certainty to judge the effects on man by analogy from experiments on animals render the problem an exceptionally severe one and we look forward to hearing what Dr Collings has to tell us in this connection, and to the paper on legal aspects of the uses of synthetic sweeteners which is to be contributed by Dr Elias.

A symposium dealing with sweetness and sweeteners must take account also of a wide range of problems which arise in connection with the use of permitted sweetening agents whether natural or artificial. Some of these arise because the substance used to provide the sweet taste fulfils functions far beyond that of mere sweetening. Sucrose, for instance, in many of the food preparations in which it is used plays an important part in giving the necessary texture to the food preparation to render it attractive and palatable. How far can the ideal texture be retained if some other sweetener with entirely different physical and chemical properties is used as a partial or complete substitute? It may then be necessary to add other substances, natural or artificial to make an acceptable article of food. We then come within range of the problems of economics in the industries involved in the production of sweetening agents, not excluding sucrose itself, and we can expect to hear something about these in the paper by Dr Selby on the use of cereal-based sweeteners. Here we are confronted with a range of problems so varied and so wide that they would warrant a whole symposium of their own. The controlled acidic and enzymic transformation of starch into mixtures of glucose, oligosaccharides and dextrins and the availability of the resulting glucose syrup as a substitute for sucrose offers a fascinating field of study both scientific and economic.

I have therefore no doubt that the present is an appropriate time for a full scale discussion on all aspects of the problems involved in sweetness and sweeteners and I look forward eagerly to hearing the comprehensive set of papers included in the forthcoming four sessions.

The Balance of Natural and Synthetic Sweeteners in Food

L. F. GREEN

Beecham Products, Research and Development Department, Brentford, Middlesex, England

ABSTRACT

The contrast between 'natural' and 'synthetic' sweeteners is shown to be diminishing. The flavour of sweeteners other than their sweetness is considered. Food formulators must be aware of metabolism of sweeteners. The relationship of chemical structure and sweetness is not known. Toxicity of sweeteners may not be limited to the so-called synthetics. Sensation of sweetness may depend on synergists and other taste factors; cost must also play its part. Legislation controls choice. Four problems seem to emerge, but formulation expertise will always be needed.

In this opening paper, I do not propose to go into any great detail about the character of the sweeteners I shall mention, for others will be able to do this far more competently than I. It *is* important, however, to define what I mean by the words 'natural' and 'synthetic'.

NATURAL SWEETENERS

In the context of this talk, I shall use the term 'natural sweeteners' to mean those sweeteners which occur in nature and which may or may not be extracted for commercial use. Modern understanding of chemical structures has enabled some of them to be synthesised and these are no longer extracted from nature, so that, straight away, it must be admitted that the distinction between 'natural' and 'synthetic' sweeteners is becoming increasingly nebulous.

Illustrative of what I am defining as 'natural' sweeteners are sucrose ($C_{12}H_{22}O_{11}$) extracted and refined from the sugar-cane or sugar-beet; maltose ($C_{12}H_{22}O_{11}.H_2O$) made by the action of the

Sucrose

β-D-Fructofuranosyl α-D-Glucopyranoside

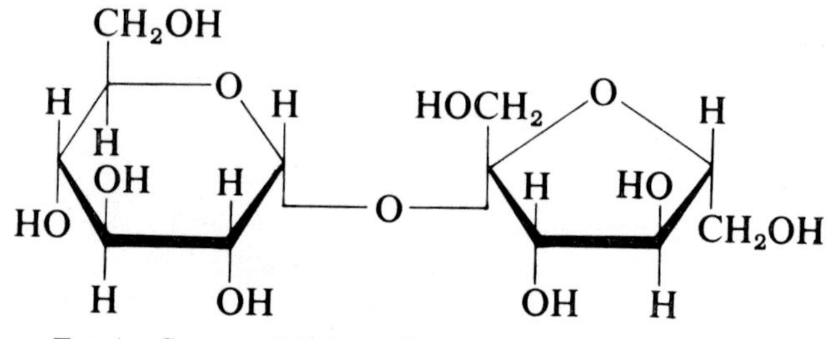

FIG. 1. Sucrose, β-D-fructofuranosyl α-D-glucopyranoside.[4]

enzyme, maltase, on starch; glucose ($C_6H_{12}O_6$) found in large quantities in grapes—hence the old-fashioned name 'grape sugar'—manufactured by the hydrolysis of starch; fructose or laevulose ($C_6H_{12}O_6$) which occurs in a large number of fruits and in honey but is, today, manufactured by synthetic methods; lactose ($C_{12}H_{22}O_{11}$) which occurs in the milk of all mammals and is prepared pure from cow's milk; trehalose ($C_{12}H_{22}O_{11}.2H_2O$), recently identified as the 'blood-sugar' of insects[1] and found in trehala manna, is a common constituent of fungi[2] such as *Amanita muscaria*, and, most importantly, in yeast[3]; and glucose syrup manufactured by the acidic or enzymic hydrolysis of various starches, usually maize in Great Britain. Out of this list, the last mentioned is notable because, although a mixture of many carbohydrates, simple

Maltose

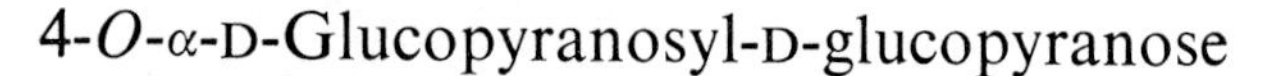

4-*O*-α-D-Glucopyranosyl-D-glucopyranose

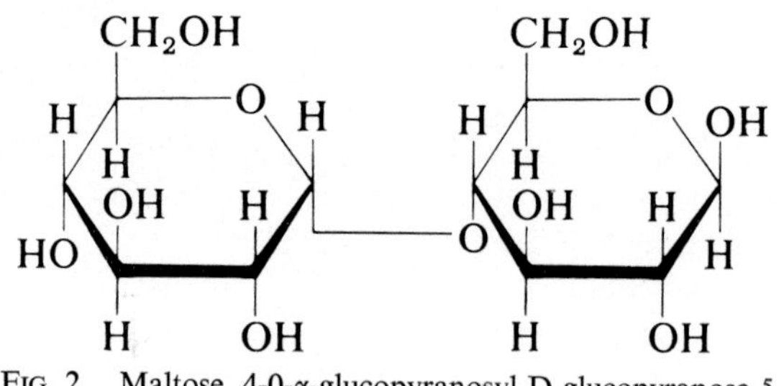

FIG. 2. Maltose, 4-0-α-glucopyranosyl-D-glucopyranose.[5]

Lactose

4-*O*-β-D-Galactopyranosyl-D-glucopyranose

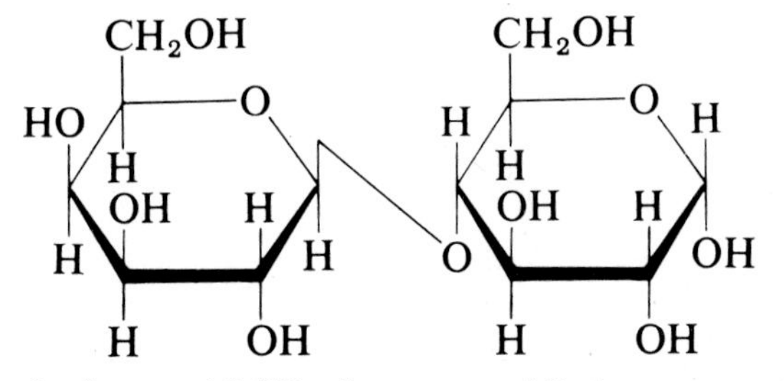

FIG. 3. Lactose, 4-0-β-D-galactopyranosyl-D-glucopyranose.[6]

and complex, it is converted solely to glucose by the body's metabolic processes.

It will have been noticed that I have shown four with the general formula $C_{12}H_{22}O_{11}$; maltose has one water molecule and trehalose has two. It is interesting to compare the structures of these four (Figs. 1–4).

α,α-Trehalose

α-D-Glucopyranosyl α-D-Glucopyranoside

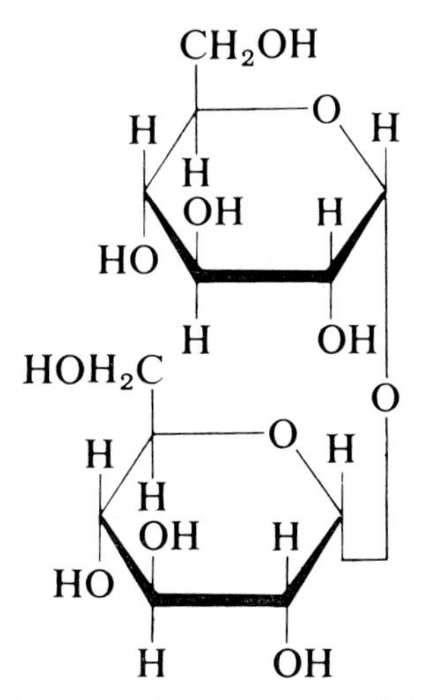

FIG. 4. α,α-Trehalose, α-D-glucopyranosyl α-D-glucopyranoside.[7]

Each of these four possesses a different degree of sweetness:

Sucrose	=	100[8]
Maltose	=	33[8]
Lactose	=	16[8]
Trehalose	=	10–15[9]

and one of the points which I hope may be brought out later in this symposium are possible reasons for these differences.

D-Glucose

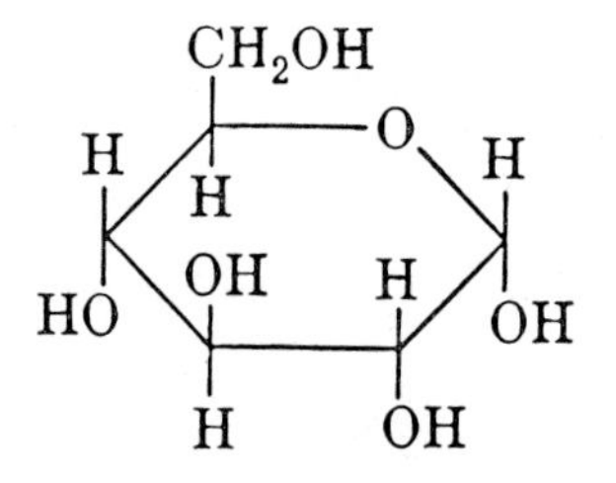

α-D-Glucopyranose

FIG. 5. D-Glucose, α-D-glucopyranose.[10]

D-Fructose

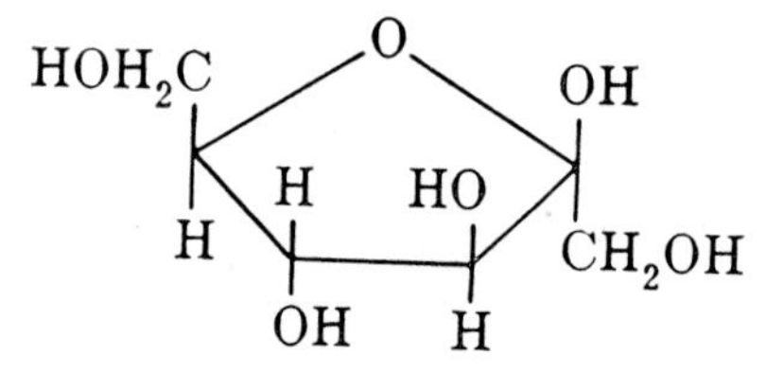

β(?)-D-Fructofuranose

FIG. 6. D-fructose, β-D-fructofuranose.[11]

Glucose and fructose are also of the same general formula, $C_6H_{12}O_6$, but again differ structurally (Figs. 5 and 6) with sweetnesses (sucrose 100) as follows:

glucose (monohydrate)	=	67[8]
glucose (anhydrous)	=	74[8]
fructose	=	110[8]

There is, then, a degree of choice for the food formulator from the various 'natural' sweeteners. However, not only are they sweet, but they possess a flavour of their own. There are three aspects to be considered: there is the flavour contribution which is made by any impurities, *i.e.* substances other than the pure sweetener. An illustration of this is the 'treacly' flavour of pre-refinery sucrose. Secondly, there is the flavour contribution made by the sweetener itself. For instance, there can be no doubt about the difference, not only in sweetness, but in flavour between maltose and sucrose. This is an aspect of flavour difference which is concomitant with the sweetness difference and which must receive attention during formulation of a new product. Lastly, there is the sensation given to the palate by the physical character of the sweetener. Many of you will be aware of the 'cool' sensation of, say, lactose. Some of these effects are lost in some formulations but they are worth mentioning in passing. It might be a rewarding study to attempt to determine whether the flavour which characterises a pure sweetener has anything to do with its relative sweetness.

The modern food formulator must also at least be aware of the metabolic differences between so-called 'natural sweeteners', for instance:

$$\text{Sucrose} \xrightarrow{\text{invertase}} \text{glucose} + \text{fructose}$$

$$\text{Lactose} \xrightarrow{\text{lactase}} \text{glucose} + \text{galactose}$$

others yield only one simpler sugar,

$$\text{Trehalose} \xrightarrow{\text{trehalase}} 2\ \text{glucose}$$

$$\text{Maltose} \xrightarrow{\text{maltase}} 2\ \text{glucose}$$

Modern technology has provided us already with many varieties of glucose syrup, and these are characterised somewhat loosely by dextrose equivalent (DE) and, more recently, maltose equivalent (ME). A few examples of these[12] are given in Table 1.

One of the valuable characteristics of these syrups is that they are rapidly metabolised to glucose only, thus providing a readily available source of this sugar which is so important to the vital processes.

TABLE 1

Composition of various glucose syrups

DE	Dextrose %	Maltose %	Maltotriose %	Higher sugars
25	4	7	11	78
32	6	45	12	37
42	21	15	10	54
55	32	19	13	36
63	38	35	12	15
90	88	4	3	5

A few instances of 'natural' sweeteners other than carbohydrates are glycyrrhizin ($C_{42}H_{62}O_{16}$) with a sweetness of 60 (sucrose = 100)[8]; stevioside or rebaudin (probably a mixture of several substances),[13] sweetness about 150 [16]; *synsepalum dulcificum*,[14] which, 'when kept in the mouth for a few minutes, makes subsequently eaten acidic foods taste very sweet',[15] and sorbitol ($C_6H_{14}O_6$), with sweetness 54.[16] The serendipity berry (*Discoreophyllum cumminsii*) is said to have an intense sweetness so great that the Africans do not use it.[17]

SYNTHETIC SWEETENERS

The term 'synthetic' has unfortunate connotations for many people, but to a gathering such as this, it merely implies what the dictionary describes[18] as a compound which is built up 'of separate elements', or an 'artificial production' of a compound from its constituents as opposed to its extraction from a plant. Two well-known examples of synthetic sweeteners are saccharin and cyclamic acid (each also supplied as the sodium and calcium salts). Saccharin is known chemically as 2,3-dihydro-3-oxobenzylsulphonazole (Fig. 7).

Cyclamic acid, or N-cyclohexylsulphamic acid has the formula as in Fig. 7.

The usually quoted sweetness figures[16] relative to sucrose (100) for these two products are:

Saccharin = 550
Cyclamate = 30

It is very difficult to see any structural relationship between the two.

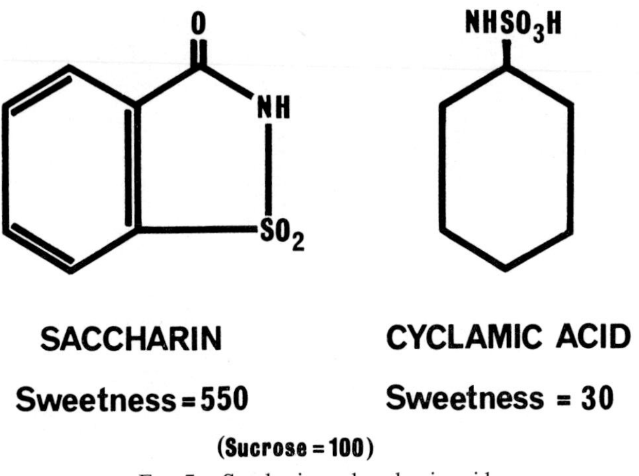

FIG. 7. Saccharin and cyclamic acid.

The ability of the taste cells to respond to a given taste substance is said to depend on the molecular mosaic that exists on the excitatory surface and a slight change may alter the taste sensation. For instance, dulcin shown in Fig. 8 is a very sweet compound whereas p-ethoxyphenylthiourea is an extremely bitter compound[19] (Fig. 8). 'The discovery of sweet-tasting compounds has been completely accidental. It is not possible to predict with any certainty whether a new structure will taste sweet or even have a taste at all. . . . Structure–taste relationships consist merely of tasting compounds and looking

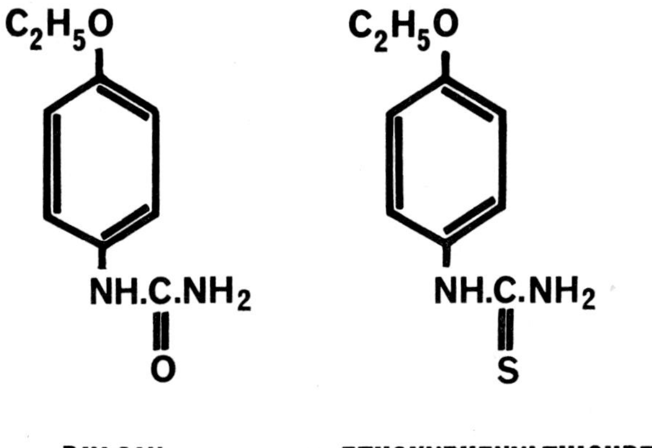

FIG. 8. Left, dulcin; right, p-ethoxyphenylthiourea.

at their structures. Correlations have been attempted but they have no predictive value outside a particular series' state Mazur, Schlatter and Goldkamp in their report of the discovery of a dipeptide sweetener, aspartyl-phenylalanine methyl ester (Fig. 9).

Such compounds give a sweetness of '100–200 times sucrose depending on concentration' and are said 'to be devoid of unpleasant aftertaste'.[20] It is interesting to note that the corresponding free carboxylic acid is completely lacking in sweetness.[21]

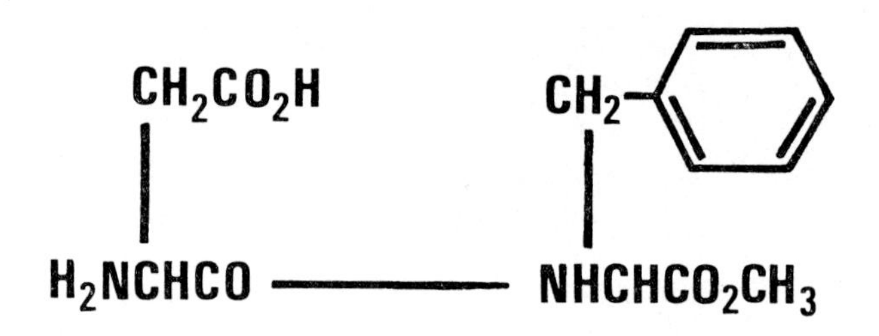

FIG. 9. L-aspartyl-phenylalanine methyl ester.

It is clear that a lot of fundamental work is called for to determine what causes the impression of sweetness to the palate, how to synthesise compounds with this property, and how so to cheapen their production that they are commercially useful. This last factor may well depend on the degree of sweetness of the compound for obviously, in general terms, the sweeter the synthetic material, the less is needed to sweeten. However, the problem does not end here for any such compound needs to be evaluated toxicologically. 'Compounds (nitro-amino alkoxybenzenes) under the general name P-4000 having 4000 times the sweetness of sucrose' have been developed 'but abandoned because they were found to produce kidney damage'.[22] Work by Dorfman and Ness[23] has demonstrated anti-androgenic activity in steviol and dihydrosteviol, the diterpinoid acids derived from stevioside. It is no news, for instance, that cyclamic acid metabolises in some individuals to cyclohexylamine which is eliminated in the urine. The latter compound is alleged to be carcinogenic and much work is now being done to clear up the many possible misconceptions about cyclamate which have caused it to be banned in America, Britain and other countries. This serves as a strong reminder that it is often considered unnecessary to put so-called 'natural' products through the same intensive screening although Birch has drawn together some of the physiological ill-effects of rare food-sugars.[24] Nevertheless, both sucrose and fructose have received recent attention as possible contributory factors to atherosclerotic conditions.[25] It emphasises also that the distinction between the terms 'natural' and 'synthetic' is somewhat arbitrary and becoming less and less distinct. For instance, as has already been noted, fructose is, today, manufactured by a synthetic method, whereas I have classed it as a 'natural' product. It surely cannot be long before many more synthetically prepared carbohydrates become available commercially and the distinction between 'natural' and 'synthetic' will become completely lost.

For the purpose of this paper, however, I am trying to make *some* distinction so as to clarify the problem which the food formulator has before him when devising a new food. He has a selection of sweeteners, some of which may be described as 'natural' and some as 'synthetic'. Each of them has a quoted sweetness, usually by comparison with sucrose. At this present point in time, he has *no* choice as regards synthetic sweeteners, for saccharin, either as such, or as the sodium or calcium salt, is the only one permitted. This has its limitations, for it is not stable to heat and will decompose at 60°C. Out of the 'natural' group he may be again somewhat limited in his choice by the nature of the product he wishes to sweeten. Each sweetener brings with it its own 'flavour', quite apart from its sweetness, and a judicious choice can sometimes be made so that this additional flavour may act as an adjunct to the overall flavour of the product.

Sweetness is one of the four primary tastes (salty, sour, bitter and sweet) but the senses of smell and touch, temperature and pain, also contribute.[26] It is usually assessed by comparison with that of sucrose in aqueous solution, and these are the figures quoted earlier, but the food formulator will find that each flavour–factor plays its part in modifying the sweetness impression on the palate. This is particularly true of the acid factor. For instance, an aqueous solution of sucrose containing 0·5% citric acid will taste a lot less sweet than the plain, aqueous solution. The relationship appears to be dependent upon 'titratable acidity' rather than pH. A change in the type of acid will not alter this relationship, but it will alter the character of the acid flavour, *e.g.* lactic acid is quite different from citric acid in flavour. But the formulator must also take into account, from another point of view, that he is working with sucrose. To produce the same sweetness with another sweetener will require, usually, an appreciably different percentage strength. This will have the effect of altering the viscosity of the solution, commonly known in soft drinks, for instance, as the 'mouth feel', which itself can modify the sweetness impression and alter the flavour contribution from the sweetener.

TABLE 2

Relative cost of sweetness

Sucrose	1
Saccharin	0·038
Cyclamate	0·32
Glucose (Dextrose)	1·4
Liquid Glucose BPC	2·7
Sorbitol	5·6
Fructose	15·0

It will be readily appreciated that this is only the beginning of the story. Foods are not just acid/sweetener solutions; they have other ingredients, all of which contribute other flavour factors. Special additives are being developed to act as synergists and some ingredients act in this way also, emphasising either the sweetness or some other factor.

Brook[27] has mentioned that 'as the concentration of saccharin in solution increases, its relative sweetness compared with sucrose appears to fall' and that 'some bitterness becomes apparent with both compounds (saccharin and cyclamate) as concentration continues to rise.' He also states that 'in citrus drinks it (cyclamate) can be 60–120 times sweeter than sucrose.'

What I am really saying, then, is that, although the food formulator must be aware of the theory and basic practical work to do with sweetness, his real task is to deal with sweetness in the context of the product he is formulating. Apparently minor changes have quite important repercussions on flavour and one of the skills of the food formulator is to be able to estimate these effects, especially when it is realised that the sweetener may contribute a major part to the cost (Table 2).

However, when either or both saccharin and cyclamate are used for sweetening, bulk is lost which may have to be compensated. This calls for one or more extra ingredients which may be more expensive than the sucrose omitted.[28]

The balance of natural and artificial sweeteners in food in Great Britain and many other countries is controlled to a certain extent by legislation. For instance, the Soft Drinks Regulations allow the following (Table 3):

TABLE 3

Mandatory sweetener contents in UK—soft drinks[29]

	Ready-to-drink			For consumption after dilution		
	Min. sugar lb/10 gal	Max. saccharin	Max. cyclamic acid	Min. sugar lb/10 gal	Max. saccharin	Max. cyclamic acid
Sweet	4·5	56	933	22·5	280	4666
Semi-sweet	2·25	56	933	11·25	280	4666

The term 'sugar', in this context, means 'any soluble carbohydrate sweetening matter' but, of course, the use of cyclamic acid and cyclamates is now forbidden.

Most of my personal experience has been with the formulation of soft drinks, but I am sure there will be an increasing interest in many different carbohydrate sweeteners, in the search for other synthetic sweeteners, and in proving these acceptable for food purposes. Advances in food technology are accompanied by advances in the understanding of metabolic processes and it is becoming more and more important for the food technologist to be well-advised on the metabolic and toxicological aspects of the ingredients he employs. There seem to be four chief problems arising: the basic factor(s) causing the sweetness sensation to the palate; the synthesising of compounds embracing that factor; the proving, metabolically and toxicologically of such compounds; and the search for possible synergists that will enhance the sweetness factor.

When these steps have been covered, the food technologist can step in with his own expertise to make good use of the materials provided.

REFERENCES

1. Karlson, P. (1965). *Introduction to Modern Biochemistry*, 2nd ed., Academic Press, New York, p. 297.
2. Pigman, W. (1957). In *The Carbohydrates*, 1st ed., Academic Press, New York, p. 507.
3. Birch, G. G. (1970). *Proc. Biochem.*, July, p. 9.
4. Pigman, W. and Horton, D. (1970). In *The Carbohydrates*, 2nd ed., Vol. IIA, Academic Press, New York, p. 101.
5. *Ibid.*, p. 107.
6. *Ibid.*, p. 104.
7. *Ibid.*, p. 106.
8. Beattie, G. B. (1940). *Flavours*, April, p. 7.
9. Birch, G. G., Cowell, N. D. and Eyton, D. (1970). *J. Fd. Tech.*, **5,** No. 3, p. 279.
10. Pigman, W. (1957). In *The Carbohydrates*, 1st ed., Academic Press, New York, p. 91.
11. *Ibid.*, p. 95.
12. Palmer, T. J. (1970). In *Glucose Syrups and Related Carbohydrates*, Ed. by Birch, Green, Coulson. Elsevier, London, pp. 25 and 27.
13. *The Merck Index* (1952). 6th ed., Merck and Co. Inc.
14. Inglett, G. E. and May, J. F. (1968). *Econ. Bot.*, **22,** p. 326.
15. *Nut. Rev.*, **28,** No. 4, p. 96.
16. Allen, R. J. L. and Brook, M. (1957). *Am. J. clin. Nut.*, **20,** p. 163.
17. Inglett, G. E. and May, J. F. (1969). *J. Fd. Sci.*, **34,** p. 408.
18. *The Concise Oxford Dictionary*, 5th ed., Clarendon Press, Oxford, p. 1313.
19. Beidler, L. M. *Flavour Research and Food Acceptance*, Rheinhold Pub. Corp., New York, p. 7.
20. Mazur, R. H., Schlatter, J. M. and Goldkamp, A. H. (1969). *J. amer. chem. Soc.*, **91**:10, p. 2684.
21. *Pat. Spec.* 1,152,977, USA.
22. Brook, M. (1970). *Dental Health*, July/Sept., p. 46.
23. Dorfman, R. I. and Ness, W. R. (1960). *Endocrinology*, **67,** p. 282.
24. Birch, G. G. (1969). *Food World*, p. 6.
25. Macdonald, I. (1970). In *Glucose Syrups and Related Carbohydrates*, Ed. by Birch, Green, Coulson. Elsevier, London, p. 86.
26. Pfaffman, C. (1958). *Flavour Research and Food Acceptance*. Rheinhold Pub. Corp., New York, p. 31.
27. Brook, M. (1970). *Dental Health*, July/Sept., p. 47.
28. *loc. cit.*
29. *The Soft Drinks Regulations* (1964), HMSO, London.

DISCUSSION

Judd: Glucose syrup and sucrose have a preservative effect in solution due to their osmotic pressure. Does this have to be taken into account when altering the ratio of synthetic sweetener to sugars?
Green: Not in soft drinks, but in jams, yes; it is also important to keep the solids content high.
Alexander: Could Mr Green please explain the basis of his relative sweetness figures? Was this experimental? We know that sweetness varies with concentration and application, but I am surprised by the rather low figure of 110 given for fructose.
Green: My figures are not based on our own experimental work; they are normally accepted figures.
Shallenberger: Crystalline fructose is about twice as sweet as sucrose. Mutarotated fructose is only slightly sweeter than sucrose.
Arnold: Fructose has very variable sweetness; it will be only 105–110 in hot beverages, but up to 200 in cold, fruit-flavoured materials. The price is now about 35 p/kg.
Pangborn: Although fructose is sweeter than sucrose at all concentrations, in distilled water solutions, these two sugars were of equivalent sweetness when compared in fruit beverages, such as peach and pear nectars [Pangborn, R. M. (1963). *J. Fd. Sci.*, **28,** p. 726].
Catlow: In view of the queries made on the dangers of high sugar intake and of cyclamate intake, would Mr Green comment on the stability of saccharin in soft drinks?
Green: What aspect would you like me to comment on? What dangers? To the best of my knowledge, instability is most unlikely. I have not heard of degradation happening in soft drinks.
Spence: To what extent are the flavour characteristics of the sugars, particularly coolness, an effect due to the dissolution of the sugar crystals or is such coolness also noted with the same sugars in solution?
Green: These are two separate effects, one due to contact on the tongue, the other due to the intrinsic flavour of the sugar. The coolness is not noted in solution.
Macdonald: Is there a standard method of measuring sweetness? Is there any physiological difference in sweet taste related to age, sex, wearing of dentures, etc?
Green: This is a very big field of enquiry. It has certainly been shown that people vary in their ability to taste sweetness. There are various methods of measuring sweetness which we shall hear about from Mr Spencer later. I am not sure whether any measurements have been done with the two sexes, but it has been shown that sweetness discrimination may diminish with age. Sweetness is best related to the final product medium, making it a very complicated problem, and I would not care to elaborate on it at this stage.
George: What examples are there at present of synergistic additives which enhance sweetness markedly?

Green: I believe there are one or two but I do not know their names. Perhaps one of my colleagues might like to comment.
Birch: Maltol and ethyl maltol are sweetness enhancers in use in the food industry today. They are both unsaturated sugar derivatives.
Spence: Would Mr Green like to comment on his remarks that the different sweeteners have flavour characteristics of their own?
Green: This is a personal opinion, but there are two points here: first, the effect of the sweetener on the drink and, second, the fact that so many also have a natural flavour of their own.
Gmunder: Is it true that the cooling effects with certain sugars and sugar derivatives may be connected with their crystal form?
Green: I do not know, but it might well be so.
Swain: Since citric and other acids affect the amount of sucrose and artificial sweeteners needed to achieve a certain degree of sweetness, do bitter substances, such as quinine in tonic water, have similar effects?
Green: Acidity most certainly reduces the sweetness effect, but I believe bitterness tends to enhance it.
Billington: As a comment on the previous question on the effect of bitterness on required sweetener concentrations: figures have been published showing that certain bitter flavours, such as grapefruit, appear to enhance the sweetening effect. In this case the sweetening power of cyclamate has been assessed as well over one hundred times that of sucrose, in contrast to the usually quoted sweetness factor of 30–50 times that of sucrose.
Swindells: Commenting upon the question of whether, like acidity, other tastes such as bitterness would also affect the relative sweetness of artificial and other sweeteners, it should also be remembered that in products like tonic water and bitter lemon, we are often seeking a different apparent sweetness to complement the bitterness. Thus we should require these drinks to appear less sweet and hence be regarded as more refreshing.

The Psychology of Sweetness

R. H. J. WATSON

Department of Nutrition, Queen Elizabeth College, University of London, Campden Hill, London, England

ABSTRACT

Human psychology is concerned with the investigation of the relationship between man and his environment as shown by his behaviour and by his subjective experiences.

The ability to experience the sensation of 'sweetness' has survived the evolutionary processes, and it is, therefore, reasonable to assume that it has performed a useful function in the development of man's selection of substances from his environment to use as food. A consideration of those environmental elements which provide sweet tastes, suggests that its usefulness lay in providing a means of detecting those substances capable of providing useful nutrients as against those which could be unsuitable.

In order for this mechanism to be effective it is necessary for it to be coupled with hedonic tone of a positive kind, i.e. *result in pleasurable experiences. In the adult human individual this relationship may be overlaid with taste preferences which have been acquired from the culture in which the individual lives, but by and large the proposition holds good. Furthermore, the culture may in fact exaggerate the relationship.*

To understand the cultural modification of the function of sweetness, it is necessary to consider the role of food in general for the individual. Man, more than any other animal, develops behaviour patterns which satisfy several needs concurrently. Thus food will not only satisfy hunger, but will also enable the individual to experience a feeling of belonging to a group—those who eat this type of food (e.g. *national food differences*) *and also satisfy feelings of social status* (*middle-class diet* versus *working-class diet*). *In this way the food habits of a group in the community tend to be self-perpetuating. The high intake of sweets and other sucrose-containing foods in this country is not due to some enhanced physiological need for sucrose, but is due to the continuation of behaviour patterns developed as part of the cultural system. The social satisfaction when added to the pleasurable experience of sweetness, makes the sweet food preference stronger still.*

In addition to the social needs already mentioned, sweets and confectionery are also used in our society as a means of enhancing personal relationships. They are used as gifts between friends, and are often used as 'rewards' for children when they have done something of which the parent wishes to approve. They will, therefore, carry with them something of the emotional aspect of the situation in which they were given. It has been suggested that factors such as these could account for some people resorting to sweet-eating in times of stress.

Psychology in addition to being concerned with general principles, is also concerned with measuring differences between individuals. It is possible to measure a number of differences related to sweetness, e.g. *sensory threshold, preference for different degrees of sweetness, preference for sweetness against other tastes, and amount of sugar consumed per day. Some of these variables can be simulated in animal experiments providing one ignores the subjective aspect and keeps firmly to an interpretation in terms of the stimulating substance,* i.e. *sucrose and saccharin and not sweetness. Experiments both animal and human have shown that individual differences in these variables show a number of interesting relationships. For example, in male human subjects a small but significant positive correlation has been found between the amount of sucrose added to beverages, and the personality characteristic of 'extroversion'.*

Technological developments have placed within man's environment a cheap substance which has the appropriate sweet sensation producing quality, coupled with a caloric content but nothing else—namely, refined sucrose. Diets have been developed which contain the sweet taste without it necessarily being coupled to nutrients such as vitamins, and thus its former function as a nutrient detector is lost. It follows, therefore, that in order to receive adequate nutrition in an environment which provides substances such as refined sucrose, it is no longer possible for man to rely solely on his taste sensitivity and preferences. He must now approach his choice of foods in a more rational manner.

From over twenty years experience of discussing psychological problems with the layman through extramural classes, and from an equally long time presenting the psychological aspects of a problem to non-psychological scientists, for example, biochemists and toxicologists, I have no doubts that there still exists considerable misunderstanding as to the nature of psychological enquiry. The view all too frequently taken is that the psychologist is someone concerned with 'the deep recesses of the mind' and that his essential piece of equipment is the couch! Let me therefore state at once that psychology covers a much wider field of enquiry than this view suggests. In fact, it is concerned with the relationship between a living organism and its environment, bearing in mind that the word environment includes other living organisms, especially those of the same species. The relationship is shown by the behaviour of the individual, and by the subjective experiences of the individual. As we shall see later, the nature of the data being considered may rule out the possibility of considering subjective aspects.

In man, the ability to experience a specific subjective state, which in our language group he comes to associate with the word 'sweetness', and which results from the stimulation of taste buds by certain substances, is the starting point of our enquiry. Here we have a clear example of the relationship between the individual and the environment—in this case the chemical substances—shown by the subjective experience. It is therefore essential to ask what function does this ability perform in the life of the individual? That it serves no function at all and is merely a vestigial function seems untenable on many counts, and it is therefore more useful to consider that the ability has survived the evolutionary process because it has real significance for the individual's survival.

The primary function of taste sensitivity lies in determining which parts of the environment contain substances suitable as food and which do not. Of the four basic taste sensations—sweet, salt, sour, bitter—the first two can be thought of as indicating the possible presence of foods, whereas the last two can be thought of as indicating possible harmful substances. At this point you may object that we frequently consume substances which are, say, bitter, for example coffee, but it must be remembered that in addition to taste selection, smell selection also plays an important role and can overlay the taste factor, and that, as we shall see later, cultural determinants also have a part to play.

It is not without significance that the first encounter with the environment in food terms made by the new-born baby results in the intake of substances which produce the sensation of sweetness. This applies to both human and artificial milks. Subsequent development of the child's food habits also confirms in general terms that sweet and salt tastes are linked to foods, and sour and bitter tastes result in the rejection of the substances. Furthermore, a consideration of the sources of the sweet tasting aspects of the environment—*e.g.* milks, ripe fruits, some vegetables—suggests that the sweet taste would be a useful 'indicator' of the presence of suitable nutrients.

However, for such a mechanism to be effective it is necessary for the taste sensation to be coupled with hedonic tone of a positive kind, that is, result in pleasurable experiences. This is generally true of the sweet and salt tastes, and the converse is true of the sour and bitter tastes. But one has to bear in mind that the hedonic tone can vary with the degree of stimulus present, a phenomenon found in all sensory modalities. In a complex pattern of sensations involving taste, smell, texture, and temperature, acceptability is accompanied by an overall positive hedonic tone, except perhaps in a few pathological cases.

In most cases, certainly in adult life, these basic mechanisms are overlaid by taste preferences which have been acquired from the culture in which the individual lives. In some instances the overlay is in fact an exaggeration of the basic relationships. In order to understand the cultural modification of the function of sweetness it is necessary to consider two aspects of human psychology which are extremely relevant to the development of food habits. The first is the fact that man in comparison with all other species has relatively the least inborn patterns of behaviour. His development relies predominantly on acquiring through learning practically all his behaviour patterns. This will apply to feeding behaviour, and therefore he will acquire from the society in which he is living the characteristic food choice patterns of the community.

The second aspect is that man more than any other species develops behaviour patterns which satisfy several needs concurrently. Of these non-nutritional needs perhaps the two most important are that of needing to belong to a group and that of defining one's status relative to one's fellows. If we consider the first of these, namely the need to belong to a group, this is satisfied within the food context by partaking of those foods which are considered as being indicative of the group, for example, national food differences. It is surprising how powerful this factor can be. Some years ago when there was a potato shortage, it was found that people were prepared to pay a very high price for potatoes in spite of the fact that there were available alternatives to provide the largely carbohydrate component of the meal, for example, rice and spaghetti. Could it be that these foods carried with them overtones of 'Indian' and 'Italian' which made them unacceptable as items in an 'English' meal?

Different foods also carry with them certain significance in terms of social status. Nutritionally the herring is, if anything, slightly superior to the salmon, but because of price the salmon carries with it a social status attribute. Similarly, fish and chips is often regarded as working class-diet, and some people erroneously think that it must be an inferior meal. In point of fact it provides a very good meal.

When these two functions, namely acquiring behaviour through learning, and the satisfaction of several needs concurrently are operating, the food habits of a group within the community tend to be self-perpetuating. The high intake of sweets and other sucrose-containing foods in this country is not due to some enhanced physiological need for sucrose, but is due to the continuation of behaviour patterns developed as part of the cultural system. These patterns were not unrelated to economic factors such as the cultivation of sugar crops within the area formally organised as the British Empire. In this context, too, one has to consider the part played by advertising in furthering and maintaining those patterns of behaviour involving the consumption of sweet-tasting foods. The satisfaction of social needs when added to the pleasurable experience of sweetness makes the sweet food preference stronger still. Just how strong the preference for sucrose-containing food or drink can become, even without social enhancement, was demonstrated in laboratory rats in the now classic experiments of Harriman,[1] in which he demonstrated that whereas naïve adrenalectomised rats showed a preference for water containing sodium chloride over water containing sucrose, animals first given a choice between these solutions when intact prefered the sucrose, and this preference was maintained after adrenalectomy in spite of the now physiological demand for large amounts of sodium.

In addition to the social needs already mentioned, sweets and confectionery are also used in our society as a means of enhancing personal relationships. This, of course, utilises the pleasurable experience which accompanies the sweet sensation. Sweets are used as gifts between friends, and are often used as rewards for children when they have done something of which the parent wishes to approve. Sweet foods will therefore carry with them something of the emotional aspect of the situation in which they were given. That is, they will tend to recreate something of the approval, the friendship and sense of belonging. It has been suggested that factors such as these could account for some people resorting to sweet-eating in times of stress.

Psychology has two faces. In addition to being concerned with general principles, it is also concerned with measuring differences between individuals. It is possible to measure a number of variables related to sweetness and to express the differences between individuals in respect of these variables in the form of a score or value. Perhaps the first one to consider is the sensory threshold. This is usually measured in terms of the concentration of sucrose or some other sweet-tasting substance in solution, which the subject will discern on 50% of the trials. There is not time to discuss the methodological problems of sensory threshold testing which will be familiar to those of you who are involved in this aspect of the problem. Another of the variables which can be measured is the preference for different degrees of sweetness. Here we find a number of different types of response. For example, there are those who prefer very little sweetness, or even none at all in certain instances, to those who prefer the very sweet sensation. Then there are those whose optimum is in the middle with a falling off of liking on either side. One can also measure the preference for sweet-tasting food against other tastes, *e.g.* salt or bland, and one can measure the amounts of sucrose consumed per day. There are many more.

Some of these variables can be simulated in animal experiments providing one ignores the subjective aspect and keeps firmly to an interpretation in terms of stimulating substance, *i.e.* sucrose and saccharin, not sweetness. This limitation is frequently forgotten and one finds researchers commenting on the human subjective sensations which they have attributed to their experimental animals! It is not possible to determine whether or not the subjective experience of the rat, such as it is, to sucrose is in any way the same as that to saccharin or cyclamate.

We ourselves have been investigating a number of these variables using both animal and human subjects. Using laboratory rats we asked the question:'to what extent do laboratory rats show a preference for one type of carbohydrate in the diet over another?' Isocaloric diets were used, the only difference being in the type of carbohydrate present (in one case sucrose, in the other case starch). Texture was made as similar as possible by using a finely powdered form of sucrose. We found that rats showed a range of individual differences from those who chose a diet in which starch was the only carbohydrate, through those with various mixtures of starch and sucrose, to those who chose sucrose as the only carbohydrate. In spite of this large range, we were able to establish a difference between males and females in that, on average, the females selected relatively more sucrose than starch compared with the males. What was perhaps more interesting was the fact that within a group of the same sex, small but significant correlations were found between preference for sucrose and lower growth rate and increased exploratory activity.[2]

It would, of course, be of interest if we could undertake comparable experiments in human subjects, but the cultural overlay is so great that interpretations in terms of physiological sexual differences would not be possible. We have found, for example, that when comparing groups of men and women in terms of the amount of sugar which is added to beverages, a considerable number of the women resolutely add none at all because they desire to stay slim. However, in the case of the men, we have been able to show a small but significant relationship between the amount of sugar added to beverages and the personality characteristic of 'extroversion'.

We have investigated the extent to which the preference between sucrose and starch in rats can be changed experimentally. It has been shown that it could be changed by varying the concentrations of the other major nutrients—fat and protein.[3] As the protein level was increased there was a shift to a preference for relatively more sucrose over starch. This was also present as fat was increased, but to a lesser extent. It was not possible to ascertain the reason for this relationship although it could have been related to the problem of establishing an overall positive hedonic tone from the mixture. The diet containing the high level of protein (casein) may have been unacceptable when mixed with starch but more acceptable when mixed with sucrose, since the sucrose could have overcome any possible adverse tastes from the casein. There is of course the equivalent in human situations where sugar is added to overcome some adverse taste, for example the tartness of acid fruits, or the bitterness of coffee. This fact has misled many weight-watchers who have erroneously assumed that 'bitter lemon' drinks cannot possibly contain sucrose.

Perhaps a more intriguing finding was that obtained when we examined the effect of differences in husbandry on the preference for sucrose over starch. Rats which were caged singly selected relatively more sucrose and less starch than did those which were caged in pairs. Furthermore, when single animals were now paired and the paired animals were separated their preferences shifted in the predicted direction. This difference was true whether the pairs were all male, all female, or male-female mixed.[4] It could be argued that for animals which had been living in groups until the experiment, the separation into single cages resulted in greater 'stress' being experienced, and that this was the factor responsible for the higher preference for sucrose. In support of this view there was additional evidence from other experiments in which, on the first day, there was a much higher preference for sucrose than on subsequent days. We have also shown[5] that experimental manipulations such as new cages, changing diets, etc. constitute 'stress'.

It is not possible to make any generalisation from this type of experiment to the human situation, except to remind you that the view has been expressed that in some individuals, at any rate, sweet eating is said to occur as a result of life stress. However this has not been shown experimentally.

We have also looked at the problem of whether early life experience of particular carbohydrates can effect subsequent preferences. Greenfield has shown that this is certainly the case in rats,[6,7] and her data suggest that there may well be a 'critical period' in infancy when the greatest effect is obtained. In addition, rats which had had access to either sucrose or starch prior to weaning showed significantly different preferences when adult, when compared to those which had been denied access until after weaning. The concept of 'critical periods' in human development is now well established. We have to ask whether or not in the human baby there exist critical periods in terms of dietary experience—for example, at what stages are particular tastes important, in our case that of sweetness. Manufacturers of baby foods, both milks and prepared meals will have an important role to play if such is the case. As you will know, at present the practice for, say, artificial milks varies from brand to brand and some require the addition of sucrose.

Finally, a few remarks concerning possible future trends would not be amiss. Sweetness, it would seem, is basically a sensation which produces pleasure and at the same time has in the past enabled man to select from his environment substances which constitute foods, not only in terms of calories, but also other essential nutrients such as vitamins. Technological developments have placed within man's environment cheap substances which are capable of producing the sensation of sweetness. In the case of sucrose it does have coupled with it calories but nothing else, whereas in the case of the artificial sweeteners even the calorie accompaniment is lacking. It is, therefore, now possible to develop edible mixtures which have the sweet taste without being coupled to nutrients such as vitamins or even to calories. Thus the former function of sweetness as a nutrient detector is becoming lost, a loss which appears to be on the increase. It follows, therefore, that man, living in an environment containing such substances as refined sucrose, saccharin, cyclamates, etc.

together with the vast range of synthetic flavourings, colourings, texturings, etc. that modern technology has made possible, can no longer rely on his taste sensitivity and preferences to select for him an adequate diet. He must now approach his choice of foods in a more rational manner, based on a knowledge of their content and his needs, rather than rely on his sensory equipment. However, unless the foods give satisfactions both physiological and psychological, the effort to produce a more adequate diet will be largely wasted.

REFERENCES

1. Harriman, A. E. (1955). *J. Nut.*, **57,** p. 271.
2. Watson, R. H. J. (1964). *Proc. Nutr. Soc.*, **23,** p. xli.
3. Shim, K. F. (1968). *The Relationship between the Composition of the Diet and the Carbohydrate Preference in the Rat*. Ph.D. Thesis, University of London.
4. Watson, R. H. J. (1966). *Proc. Nutr. Soc.*, **25,** p. xiii.
5. Steinberg, H. and Watson, R. H. J. (1960). *Nature, Lond.*, **185,** p. 615.
6. Greenfield, H. (1969). *Proc. Nutr. Soc.*, **28,** p. 19a.
7. Greenfield, H. (1970). *Early Dietary Experience of Particular Carbohydrates and Subsequent Preference in the Rat*. Ph.D. Thesis, University of London.

DISCUSSION

Bishop: In Western man, brought up on a diet of sucrose as sweetener would you expect physiological disturbance, *i.e.* withdrawal symptoms if placed on a diet with starch and artificial sweeteners? When an animal is short of salt it will lick a salt cake; if salt and sweet tastes are food pointers, will an omniverous animal search for carbohydrate when deficient?
Watson: A good point. What you are asking is whether the removal of sucrose and saccharin from the diet would result in withdrawal symptoms. I would not think so. Unless one can say that there is some particular link with the fructose part of sucrose, I think withdrawal symptoms are unlikely.
Bishop: What distinctions are there between eating for pleasure and of necessity?
Watson: Most of the satisfaction of adding sugar to a drink is the continuance of a behaviour pattern from the past. The sight of sugar granules going in, instead of one granule of saccharin, gives pleasure. But if you are just presenting saccharin without the action, then I do not think that loss of pleasure would be experienced. The comments I made were, in fact, in the human situation, and not in cattle. If you suffer salt loss due to extensive exercise then the preference pattern changes, and it was shown by Haldane that salt drinks are then extremely pleasurable. It would seem

that our taste receptivity changes. The sugar factors are the predominant ones in humans. We have got so used to certain effects and tastes that we are unable to break out of this. The social pattern is predominant.

Griffiths: I was surprised that sweetness was of the importance you suggest in selecting nutrients in primitive man—a meat and cereal eater. Would you comment?

Watson: Many people would dispute about primitive man being a meat eater; one can think in terms of fruits and berries as being sources of vitamins, and that the sweet taste is, in fact, an indirect factor. Meat eating is closely associated with odour. We are only dealing with a tiny little fragment of the total sensation and it can, in fact, be severely overlaid by cultural factors. For instance, you will find people taking bitter substances because it is the tradition.

Green: Forty-five years ago, teenage parties used sweet jellies, cakes, etc; today, teenagers prefer savoury foods. Can you comment?

Watson: Everybody has experienced this. Children are not so keen today on jellies and ice creams; they want savoury snacks. The factors which have given rise to this change are, partly the development of convenience foods by the food manufacturers, and partly, that the types of food which a community will take are not constant. From time to time they will change, but very slowly. Look at the way immigrant populations react to some of the host-country's food habits, it takes a very long time for them to adapt.

Social sub-groups have their own dishes; for instance, fish and chips were the working-man's dish, but the higher social classes, partly because of financial resources, utilised more savoury foods and less of the sweet kind. If you are wanting to put on a spread, what you try to do is to present things which give additional social status. You want to offer something unusual. With the additional cash now available, you are able to bring into the celebratory field these more expensive savoury foods.

Swindells: Commenting upon the nature of changes in food preferences, does Dr Watson feel that it is likely to be valuable to identify and then target advertising and promotion at teenagers, *i.e.* opinion leaders—always provided that the difficult job of locating the teenagers for the particular situation can be solved?

Watson: This is a difficult problem to solve. How do you identify leaders of trends? What I would suggest is we have to have a completely different approach to foods, for it is not enough to rely on taste sensitivity alone. Just to give you an example; one learns at school that a good source of vitamin C is fresh oranges, and one might assume that if one takes an orange drink, that one gets vitamin C; but there are orange-coloured drinks which do not contain vitamin C at all! How is the individual going to face up to this? It is a problem, and we have to think in terms of nutritional education and the enjoyment of food, not just as a cold experience. We must look at the psychological aspect as well as the scientific.

There is a celebrated case of the attempt to introduce maize to an underdeveloped country and everybody thought how marvellous, but after a period of years they had reverted to their previous eating habits. The thing had failed through total non-understanding of the existing cultural food habits.

Macdonald: With reference to Dr Bishop's point earlier, a colleague of mine (Dr M. Roberts) had 19 men for 14 weeks on a sucrose-free diet in the Antarctic with no withdrawal symptoms.

Swain: 1. With regard to withdrawal symptoms on removal of sucrose from the diet, it should be remembered that in 1939–45 the whole population of the country was exposed to this effect and no apparent withdrawal symptoms were noted.

2. With regard to the present taste for savouries, it should be remembered that jellies and blancmanges, which were popular 40–50 years ago, were introduced only in the latter part of the 19th century. Before that, the menus of celebratory meals or parties were mostly savoury. Present day young people are returning to the preferences of their great-great-grandfathers rather than their grandfathers and fathers.

Summers: In the experiments with rats, and preference as between sucrose and starch, has Dr Watson done anything with added flavours to the diets, *e.g.* adding what we might consider an acceptable flavour to starch and an unacceptable flavour to sugar?

Watson: No. We were not interested in looking at subjective preferences but only to determine acceptance of sucrose or starch. In an experiment where cyclamate had been used, there had been some rejection by rats.

Galitzine: What work has been done to investigate the genetic factor in the inheritance of taste preferences?

Watson: This is something which we plan to do. We are hoping to set up a selective study to begin to analyse what particular characteristics go with taste preference. This will take several years to set up.

Galitzine: Would it not be possible to extract data from the USA where there are immigrant populations from many different areas with various taste and food habits?

Watson: If you are thinking in human terms, one gets into difficulty when one is dealing with a behaviour characteristic. You are dealing with something which is very susceptible to cultural factors. The only way you can do it is by a genetic experiment. For instance, one can study identical twins in different environments to make a comparative study. One needs to look at all the environmental factors; too many cultural and racial influences render this type of research ineffective.

Young: Have you any opinion on the physiological reasons for the rise in sucrose intake in times of stress?

Watson: I would not like to guess. All I can say is that in the human field this has been suggested by various people. It has not really been put to any rigorous test, but, in the animal sphere, we do know that if the animal is placed in a situation of stress, this raises the blood sugar level. To what extent the change of dietary intake reflects stress is not known. We should like to do this.

Intense Sweetness of Natural Origin

G. E. INGLETT

Northern Utilisation Research and Development Division, Agricultural Research Service, US Department of Agriculture, Peoria, Illinois, USA

ABSTRACT

Sweetness is a fundamental gustatory response that, so far, can be measured by taste evaluations. Substances that occur in nature that possess sweet taste are primarily sugars or derivatives of carbohydrates. Sweetness variations among sugars are small compared to some of the more intensely sweet substances. A new sugar substitute, L-*aspartyl*-L-*phenylalanine methyl ester, was found to be* 100–200 *times as sweet as sucrose. Another sweetener, neohesperidin dihydrochalcone, is* 1000 *times the sweetness of sucrose.*

Nature is a logical place to start any search for a new superior intense sweetener. Many toxic substances are found in nature, so any new naturally occurring sweeteners, either of intense or ordinary sweetness, cannot arbitrarily be considered safe for human consumption.

Most research investigations of this type have involved studies of plant materials that are known to be consumed by various indigenous people. For example, an intensely sweet, naturally occurring sweetener is found in the leaves of a small shrub, Stevia rebaudiana *Bertoni, that grows wild in Paraguay. The natives use the leaves to sweeten their tea and other foods. Stevioside is the sweet principle.*

Another natural intense sweetener is glycyrrhizin. It is isolated from the licorice root (Glycyrrhiza glabra L.) *as the ammonium salt which finds some commercial applications.*

An unusual source of sweet taste is observed from a tropical fruit known as the miracle berry (Richardella dulcifica/*Schum. and Thorn.*/*Baehni*). *This berry possesses a taste-modifying substance that causes sour foods to taste sweet. After the mouth is exposed to the active principle, sour foods such as lemons, limes, grapefruit, and strawberries will taste delightfully sweet. The active principle, miraculin, is a glycoprotein. It was suggested for use in a recent patent as a sweetening agent in yoghurt, in preparations for buccal hygiene, and for vitamin C and aspirin.*

A screening programme for intense sweetness of tropical plants by Inglett in 1965 *gave several leads. An intense sweetness was found in the West*

African fruit called Serendipity Berries (Dioscoreophyllum cumminsii (*Stapf*) *Diels*). *Threshold taste response of the isolated active principle gave a sweetness value of* 1500 *times sweeter than sucrose which appears to be the sweetest naturally occurring substance known.*

Another African fruit that contained intense sweetness is called katemfe in Sierra Leone or the miraculous fruit of the Sudan (Thaumatococcus Daniellii *Benth*). *Preliminary studies have indicated a substance similar to the Serendipity Berry sweetener.*

TASTE EVALUATION

Sweetness is a fundamental gustatory response that, so far, can be measured only by taste evaluation. Substances that occur in nature that possess sweet taste are primarily sugars or derivatives of carbohydrates. The most common sugars are sucrose, dextrose, fructose, lactose, raffinose, mannose, xylose, arabinose, and galactose. Sweetness varies among these sugars from essentially nothing for raffinose to 1·7 for fructose with sucrose having a relative sweetness value of one. These are small variations in relative sweetness compared to some of the more intensely sweet substances shown in Table 1.

The sweet compounds listed illustrate only a few of many sweet substances. The intensity of their sweetness can rise far above that of sucrose; for example, 1-*n*-propoxy-2-amino-4-nitrobenzene (P-4000) is 4000 times sweeter than sucrose. A new sugar substitute *L*-aspartyl-*L*-phenylalanine methyl ester, was found to be 100–200 times as sweet as sucrose.[2] It was synthesised from two non-sweet

TABLE 1

Sweetness of various substances

Substance	Relative sweetness,[1] weight basis
Sucrose	1
Cyclohexylsulfamate (sodium cyclamate)	30
L-Aspartyl-*L*-phenylalanine methyl ester	100
2 ,3-Dihydro-3-oxobenzisosulfonazole (saccharin)	350
Neohesperidin dihydrochalcone	1000
Perillaldehyde *anti*oxime (Perillartine)	2000
1-*n*-Propoxy-2-amino-4-nitrobenzene (P-4000)	4000

amino acids, *L*-aspartic acid and *L*-phenylalanine. This dipeptide sweetener was reported to have taste characteristics more similar to sucrose than a cyclamate-saccharin mixture.

Neohesperidin dihydrochalcone is rated at 1000 times the sweetness of sucrose. It results from the hydrogenation of the naturally occurring flavonoid, neohesperidin.[3,4] Although neohesperidin dihydrochalcone has a good sweet taste, the aftertaste can be described as a menthol-like cooling effect or licorice-like.[5] Sweetener intensity is an important measure of economic value since the cost per sweetness-pound must be one consideration of a new sweetener's merit. Sweetness intensity is also an index of the quantity to be consumed by humans which would, of course, be less for the more intense sweetener. This could be an important health consideration for some substances.

Another important factor in sweetness evaluation is its quality. For many sweet substances, the sweet quality, which is very subjective, is not adequately investigated. Sweetness quality is measured generally by comparison with sucrose. Sucrose causes a rapid sweet taste impact followed by a sharp cut-off. Most, if not all, of the synthetic sweeteners are judged deficient in sweet-taste response by comparison to sucrose. Moreover, sucrose has 'bodying' or viscosity effects that give texture factors or smooth mouth-feeling not contributed by the synthetics. For any new commercial sweetener, sweetness quality as well as intensity must be an important consideration.

Besides sweetness intensity and quality, any new commercial sweetener must be harmless to human health. Of the many sweet substances known, only a few of the intense sweeteners can be considered safe for human consumption. Saccharin, at the present time, is the only synthetic sweetener allowed in American foods. Cyclamate was used in foods until banned in 1970 by the US Food and Drug Administration.

Nature is a logical place to start any search for a new, superior, intense sweetener. The search would be expected to be difficult, expensive, and may require considerable time and basic research. Of course, nature produces many toxic substances, so any new naturally occurring sweeteners, either of intense or ordinary sweetness, cannot arbitrarily be considered safe for human consumption. A successful venture for an intense sweetener probably would locate a natural material having intense sweetness, isolate the sweetener principle, evaluate the sweetness properties, characterise the substance, conduct toxicology studies, synthesise the compound and, perhaps, some of its related analogues. A successful sweetener will be non-toxic, taste like or very similar to sucrose, and give a sweetness intensity sufficiently high so that it can be produced economically or isolated from cultivated plants at competitive costs per sweetness-pound.

STEVIA REBAUDIANA BERTONI

Most research investigations of this type have involved studies of plant materials that are known to be consumed by various indigenous people. For example, an intensely sweet, naturally occurring substance is found in the leaves of a small shrub, *Stevia rebaudiana* Bertoni, that grows wild in Paraguay. Natives use the leaves to sweeten their tea and other foods. Stevioside, the sweet principle, is obtained in a 6% yield by ethanol extraction of the dried leaves. It is 300 times sweeter than sucrose. Stevioside[6] is a steroid glycoside with a sophorose sugar. The detailed structure of stevioside was established in 1963 by Mosettig *et al.*[7] It is a diterpenoic acid esterified to one glucose unit and combined in glucosidic linkage with sophorose, a disaccharide of *D*-glucose. The aglycone, steviol—a steroid, has some interesting biological properties.

GLYCYRRHIZIN

Another natural intense sweetener, glycyrrhizin, in the form of licorice root (*Glycyrrhiza glabra L.*) extracts has been used by pharmacists for years as a flavouring agent to mask unpleasant tasting drugs. The sweetener is present in licorice root as the calcium and potassium salt of glycyrrhizic acid. It is isolated from the root as the ammonium salt which finds some commercial applications. Glycyrrhizin is a saponin of the corresponding glycyrrhetinic acid with an attached disaccharide moiety. The structure of the sugar moiety was confirmed by periodate oxidation of glycyrrhizic acid and by degradation of trimethyl glycyrrhizate pentamethyl ether.[8] Its wider commercial application may be limited by its physiological activity.[9] The triterpenoid glycoside appears to be about 50–100 times sweeter than sucrose with some undesirable lingering taste qualities.

RICHARDELLA DULCIFICA

An unusual source of sweet taste is present in a tropical fruit known as the miracle berry (*Richardella dulcifica*/Schum. and Thorn./ Baehni). This berry possesses a taste-modifying substance that causes sour foods to taste sweet. After the mouth is exposed to the active principle, sour foods such as lemons, limes, grapefruit, rhubarb, and strawberries will taste delightfully sweet. Even dilute organic and mineral acids will induce a sweet taste. The berries are chewed by West Africans for their sweetening effect on some sour foods.

The quality of the sweetness induced by this taste modifier is very similar to sucrose. The intensity of the sweetness depends upon the concentration of active principle, the proton source, and individual gustatory response. Regardless of concentrations and sensory factors, the maximum sweetness intensity appears to be no more than obtained from a high sucrose concentration. The disadvantage of this material as a sweetener under the present state of knowledge is its lingering effect. Sweetness response can be reactivated when protons are resupplied and the principle apparently remains active. The sweetness effect can be repeated for perhaps as long as two hours.

Isolation studies of the active principle of this substance[10] established its water insolubility and lability. It was suggested that the active principle could be a glycoprotein based on solubility behaviour and denaturation by heat, acid and base. Recently, Brouwer *et al.*[11] and Kurihara and Beidler,[12] working independently, isolated the active principle. Solubilisation of the active principle, called miraculin by the Unilever group, was accomplished by using extraction aids such as salmine and a naturally occurring polyamine spermine (*N,N'*-bis/3-aminopropyl/-1,4-diaminobutane). Miraculin was purified by ammonium sulphate fractionation and gel filtration on Sephadex G-50 and Sephadex G-25. The active principle was called taste-modifying protein by Kurihara and Beidler,[12] who solubilised the active principle by extraction with pH 10 buffer for 1 minute. Purification was accomplished by Sephadex chromatography.

The active principle was identified as a glycoprotein, based on the loss of activity on treatment with proteolytic enzymes, on the positive test for sugars before and after electrophoresis on polyacrylamide gel, and by the yields of amino acids and sugars found after acid hydrolysis. Miraculin has a molecular weight of 42,000 $\pm$ 3000. As more information is obtained regarding its structure, useful data regarding the mechanism of taste perception is expected to follow. Miraculin was suggested in a recent patent[13] for use as a sweetening agent in yoghurt, in preparations for buccal hygiene, and for vitamin C and aspirin.

FIG. 1. Serendipity Berries: the intense sweet fruit of *Dioscoreophyllum cumminsii.*

SERENDIPITY BERRIES

A screening programme for intense sweetness of tropical plants[10] gave several leads that suggested the need for further evaluation and research to determine their economic importance.[14] A very intense sweetness was found in the West African fruit called Serendipity Berries.[15] Botanically, they are the fruit of *Dioscoreophyllum cumminsii* (Stapf) Diels. The fruit is not commonly cultivated or used by the natives of Nigeria because of its intense sweetness; however, it is eaten in the Belgian Congo. The berries are light red, approximately 0·5 inch in diameter, and grow in grapelike clusters with a variation of several to 100 berries in each bunch. These berries are shown in Fig. 1.

Chromatography of water extracts of the berry on Sephadex G-50 and G-200 indicated that the sweetener was bound to protein. Degradation of the fruit extract with bromelain, a proteolytic enzyme, yielded a substance of lower molecular weight (about 20,000) with intense sweetness of excellent quality. Functional group tests indicated that this substance was not proteinaceous but an apparent polysaccharide. Threshold taste response gave a sweetness

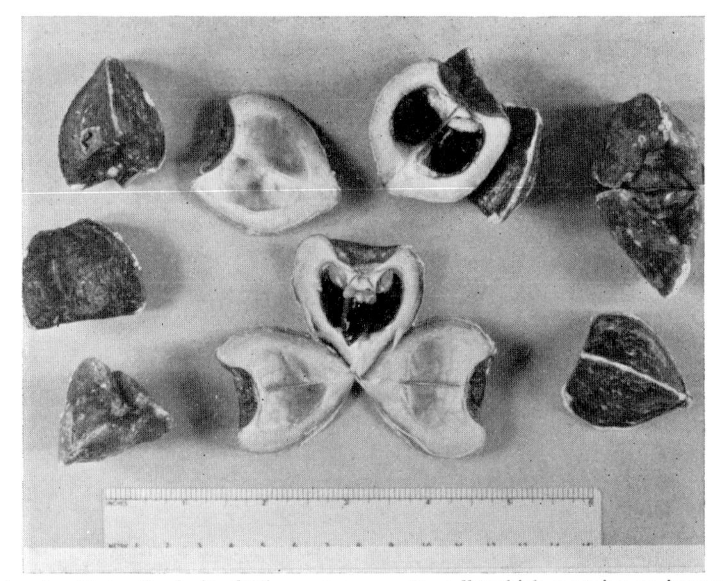

FIG. 2. Katemfe: fruit of *Thaumatococcus Daniellii* which contains an intense sweetness in the gum coating around the seeds.

value 1500 times sweeter than sucrose which appears to be the sweetest naturally occurring substance known.[15]

KATEMFE

The systematic plant study of Inglett and May[14] revealed another African fruit that contained an intense sweetener. The fruit is called katemfe in Sierra Leone or the miraculous fruit of the Sudan. Botanically, the plant is *Thaumatococcus Daniellii* Benth. The fruit contains three large black seeds surrounded by a transparent jelly and a light yellow aril at the base of each seed (Fig. 2). The mucilaginous material around the seeds is intensely sweet and causes other foods to taste sweet. The seeds were observed in trading canoes in West Africa as early as 1839, and were reported to be used to sweeten bread, fruits, palm wine, and tea. Preliminary studies have indicated a substance similar to the Serendipity Berry sweetener.

There are probably other undiscovered intensely sweet-tasting substances in nature and new sweeteners that may be synthesised. The success of any sweetener programme will depend on obtaining a superior sweetener that has an advantageous economic cost per sweetness-pound, non-toxic properties, and a taste comparable or similar to sucrose.

REFERENCES

1. Inglett, G. E. (1971). *Recent Sweetener Research*, 2nd ed., Botanicals, PO Box 3034, Peoria, Illinois 61614, USA.
2. Mazur, R. H., Schlatter, J. M. and Goldkamp, A. H. (1969). *J. Am. chem. Soc.*, **91,** p. 2684.
3. Horowitz, R. M. and Gentili, B. (1969). *J. agr. Fd. Chem.*, **17,** p. 696.
4. Krbechek, L., Inglett, G. E., Holik, M., Dowling, B., Wagner, R. and Riter, R. (1968). *J. agr. Fd. Chem.*, **16,** p. 108.
5. Inglett, G. E., Krbechek, L., Dowling, B., and Wagner, R. (1969). *Fd. Res.*, **34,** p. 101.
6. Vis, E. and Fletcher, H. G., Jr. (1956). *J. Am. chem. Soc.*, **78,** p. 4709.
7. Mosettig, E., Beglinger, U., Dolder, F., Lichti, H., Quitt, P. and Waters, J. A. (1963). *J. Am. chem. Soc.*, **85,** p. 2305.
8. Lythgoe, B. and Trippett, S. (1950). *J. chem. Soc.*, p. 1983.
9. Murav'ev, I. A. and Ponomarev, V. D. (1962). *Med. Prom. SSSR* 16, No. 8, 11. CA **58,** p. 4371 (1963).
10. Inglett, G. E., Dowling, B., Albrecht, J. J. and Hoglan, F. A. (1965). *J. agr. Fd. Chem.*, **13,** p. 284.
11. Brouwer, J. N., Van DerWel, H., Francke, A. and Henning, G. J. (1968). *Nature, Lond.*, **220,** p. 373.
12. Kurihara, K. and Beidler, L. M. (1968). *Science, N.Y.* **161,** p. 1241.

13. Henning, G. J., Brouwer, J. N. and Van DerWel, H. (1969). *Belgium patent*, 737,042.
14. Inglett, G. E. and May, J. F. (1968). *Econ. Bot.*, **22,** p. 326.
15. Inglett, G. E. and May, J. F. (1969). *J. Fd. Sci.*, **34,** p. 408.

DISCUSSION

Swain: The research which you have carried out so far is very interesting and I hope you are continuing this search for natural sweetness. Has any attempt been made to look systematically and to examine closely related plants to those known to contain sweetness, such as *Dioscoreophyllum Cumminsii,* for sweetness or, alternatively, to examine the field notes of botanical collectors which are kept in herbaria?

Inglett: These records are considerably deficient in that they do not describe the intensity of sweetness or the quality of sweetness, and the difficulty, after this, is searching for the fruit, which adds considerably to the expense. Very few people in industry, or otherwise, are in a position to finance and to support a research programme in this area, but my own aim is to carry out a programme of this sort.

Once a source of intense sweetness is located, the difficulty is to find out what is the sweetness principle, and whether it can be evaluated and characterised. The sweetness principles of the miracle berry and the serendipity berry are both of high molecular weight which devalues their industrial potential considerably and increases the difficulties of the problem. That is not to say that one cannot extract some small fraction of a large molecule, which may then be made commercially available. This does not say anything about the toxicity. It all adds up to an extremely complicated and expensive programme.

Beidler: Are any laboratories now studying the characterisation of the active material in the serendipity berry?

Inglett: Yes, but I am not sure whether I am at liberty to say, because it was obtained indirectly, and I am not sure that they would want it known, though I *can* say that its molecular weight is greater than 20,000.

Cunningham: With reference to amino acids and derivatives (*e.g.* aspartic acid, phenyl alanine, methyl ester), the fermentation industry (now turning its interest to the food industry) could be in a position to produce quantities of the dipeptide derivative, but economic consideration may rule this out. Has any work been carried out on the enhancing of quality/sweetening power of aspartic acid, say, by using nucleotide derivatives such as has been accomplished on other amino acids, *e.g.* the enhancing of the flavour of monosodium glutamate using guanylic acid, etc.

Inglett: You *can* enhance the flavour of monosodium glutamate but, as you say, the question is, are such enhancers commercially feasible? I would agree with your comment that at the present time they are not economical; costs have been worked out by several large companies. There are naturally occuring diamino acids which are intensely sweet which may have possibilities, but amino acids are additives, therefore standards would have to be changed before they could be used in foods.

Rostagno: Could you give details on physiological side-effects of glycyrrhizin, studied by the Russians?

Inglett: Later, I can give you the Russian Laboratory report on this.

Noble: Stevioside, glycyrrhizin and neohesperidin dihydrochalcone all contain a 2—0 linkage between the two monosaccharide units in the disaccharide moiety. Do you regard this fact as being important? I believe you also submitted a sample to Dr Amoore; did he have any success in determining whether or not this factor has any importance?

Inglett: I gave lectures on this six years ago, but since then I have come to think it has no meaning. You can find many 2—0 linked components but I can find no basis for believing that it is the exclusive requirement for sweetness.

Plaskett: First, with regard to the very interesting glycoprotein sweeteners mentioned in the latter part of your talk, are your indices of sweetness recorded on a weight basis or on a molar basis?

Second, you spoke as if the high molecular weight of these glycoproteins automatically makes them valueless for commercial use, unless they can be split and fractionated. Is that your view, or is there any potential for commercial application irrespective of high molecular weight?

Third, do you know of any economic evaluations of agronomic production of these glycoprotein sweeteners? Is it likely that plantations for their production could be economically viable?

Inglett: I believe that the more scientific approach is on a molar basis, but present work has been reported on a weight basis. The synthesis of high molecular weight compounds is so intricate that it is out of the question in terms of commercial possibility. There is the other possibility, that agronomic research might produce, in quantity, a material like sucrose as long as it were stable in processing. There is no reason to suspect or believe that the plantation approach for something like the serendipity berry could not be undertaken. People are growing it in Malaysia. It grows like grapes, so it *is* a possibility.

Oduro-Yeboah: Do you test both the purified form of the sweetener and the crude material for toxicity? In my opinion both forms of the sweetener should be tested in order to determine whether the extraction and the purification processes in any way contribute to any possible toxic effects of the purified material.

Inglett: This is always a problem in working with these materials and that is why one always tries to choose materials being consumed by some indigenous group. This does not mean, necessarily, that it is non-toxic. We could feed it in large quantities to rats or mice, but, of course, this is a 'do or die' situation. If long term-effects were present, these would not show up, but if the tests proved the compound to be extremely toxic, the experiment should stop unless there were possibilities that fractionation would be of use.

The Theory of Sweetness

R. S. SHALLENBERGER

New York State Agricultural Experiment Station,
Cornell University, Geneva, New York, USA

ABSTRACT

Although long known as a 'chemical sense', the initial chemistry of 'sweetness' has been difficult to pin down. Starting with the anomalous varying sweetness of the sugars, a model was constructed, based primarily on hydrogen-bond theory and conformational parameters, which explained varying sugar sweetness. The model subsequently was expanded to account for the sweetness of amino acids, 'synthetic' sweetening agents, salts of lead and beryllium, and other compounds. In essence, the initial chemistry of 'sweetness' is thought to be a simultaneous double intermolecular hydrogen bonding phenomenon with a specific stereochemical requirement.

INTRODUCTION

The recent theory[1–3] of the chemical basis of sweet taste had, as its origin, the rationalisation of the varying and seemingly anomalous sweetness of sugars. Since I sense, at this meeting, a rather general interest in the varying sweetness of the sugars, I would like to demonstrate the problems of sugar sweetness that we recognised, and how they were resolved. Finally, I would like to illustrate how resolution of this problem led to a general model for the sweet taste of a great variety of compounds, and how this, too, led to a model for the initial chemistry of the sensation of sweetness.

FINE STRUCTURE OF SUGARS

The approximate sweetness and the structure of the readily available hexoses and the ketose D-fructose are shown in Table 1.

TABLE 1

The favoured structure of selected sugars and their approximate sweetness (sucrose = 100)

Sugar	Structure	Conformation	Sweetness
β-D-Fructose	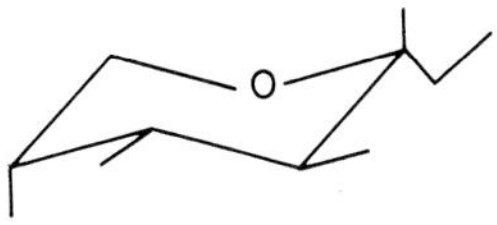	1C	180 115
α-D-Glucose	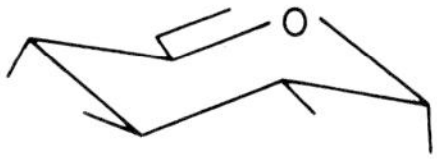	1C	74 74
α-D-Galactose	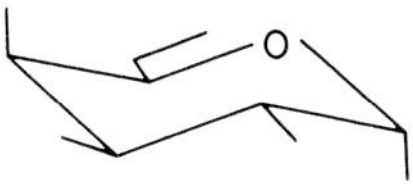	1C	32
α-D-Mannose	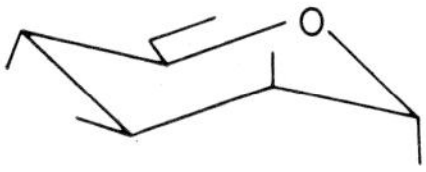	1C	32

The structure of the pyranose sugars shown are the favoured ring conformations in each case. Also, in each case, each sugar glycol unit has the favoured *gauche* or *staggered* conformation which means that in all cases, the sugar OH groups are equidistant.

Thus far, we have no basis for explaining the differences in sweet taste shown.

If, however, we assume that the glycol moiety in each sugar is the repeating unit responsible for sweet taste, possible reasons for varying sweetness begin to appear.

For fructose, the anomeric OH proton is the most acidic of the lot, and the OH group of the hydroxy methylene group is unique in that it has free rotation. Another unique feature of the fructose pyranose ring is the ring methylene carbon atom giving this sugar an element of lipoid solubility.

When crystalline β-D-fructose is dissolved in water, it mutarotates to establish an equilibrium with its β-D-furanose tautomer (Fig. 1). The former is present to the extent of 68%, the latter 32%. Since

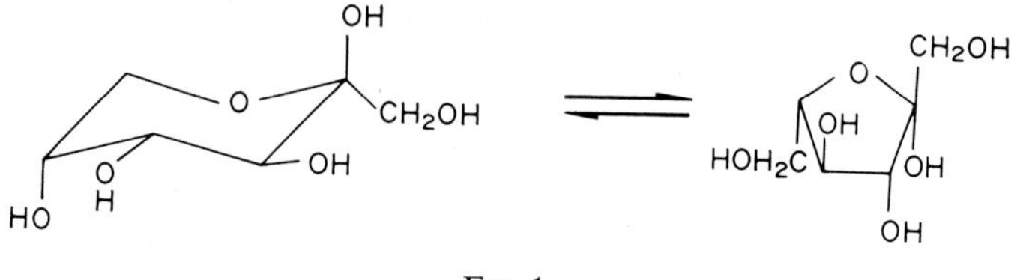

FIG. 1.

(0·68) (180) = 122, we concluded that the furanose form possesses very little sweetness. Interestingly, the vicinal OH groups are now either nearly eclipsed, or in the *anti* conformation. In the former case, they are capable of forming a strong intramolecular hydrogen bond. In the latter case, they are incapable of such bonding.

The glucoses were particularly perplexing. Since freshly prepared solutions of α-D-glucose are sweeter than mutarotated solutions,[4]

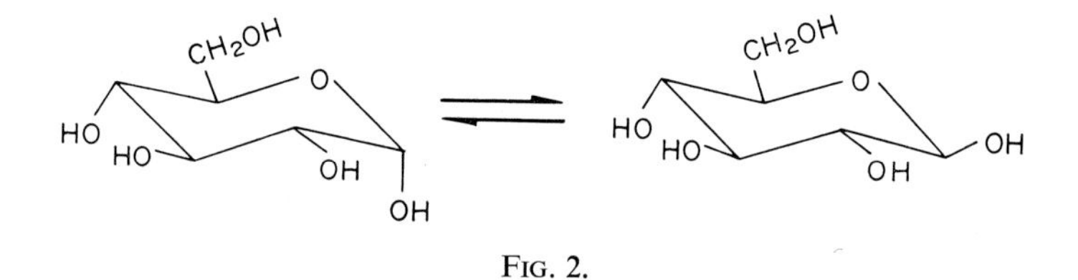

FIG. 2.

it seemed obvious that merely describing the mutarotation of this sugar as a change from the α to the β pyranosidic form (Fig. 2) with 64% β-D-pyranose form and 36% α-D-pyranose form present was not adequate to explain decreased sweetness. However, if the β-D-pyranose form persisted in an unknown amount of the chair conformation (Fig. 3), the steric disposition of vicinal OH groups would be vastly altered, from the *gauche* to the *anti* conformation. To test if vicinal OH groups in the *anti* conformation would be unable to elicit sweet

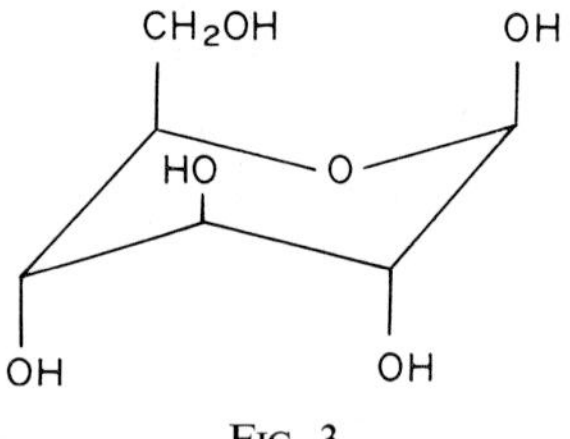

FIG. 3.

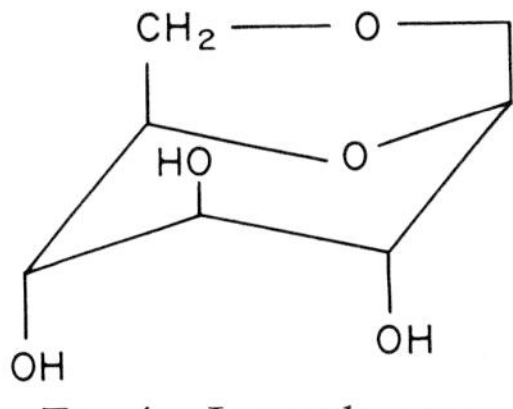

FIG. 4. Laevoglucosan.

taste we prepared laevoglucosan[5] (Fig. 4), and found that it was devoid of any taste. Parenthetically, there is Nuclear Magnetic Resonance (NMR) and Optical Rotation (OR) data which indicate an unknown amount of β-D-glucopyranose in the alternate conformation in water solution, presumably stabilised by the formation of several hydrogen bonds.

The next sugars that were considered were α-D-mannose and α-D-galactose. These hexoses are merely diastereoisomeric with glucose, and are the carbon No. 2 and 4 epimers, respectively. Thus, the OH substituents at these positions are axial. The key as to why these two sugars would be only one-half as sweet as glucose came from studies on tetrahydropyrans at the University of Birmingham.[6] That school of investigators observed that the axial OH substituents on galactose and mannose were sterically disposed to bond the ring oxygen atom.

GLYCOL CONFORMATION

As a consequence of delving into the fine structures of the sugars, two possible reasons for varying sugar sweetness evolved. In one case when the sugar glycol unit could assume a conformation wherein the vicinal OH groups eclipsed, so they could hydrogen bond intramolecularly, then sugar sweetness was diminished. On the other hand, if one OH group of a glycol pair was able to bond elsewhere in the molecule, as for example galactose and mannose, that bonded OH group would reduce the number of glycol units available to elicit sweet taste. To test this first case, the sweetness of galactose *versus* glucose was compared at different temperatures. It was found that both glucose and galactose sweetness increased with increasing temperature, but the relative sweetness of galactose increased *twice* as fast. In plotting the reciprocal of the sweetness score *versus* calculated increase in free OH, we obtained the result shown in

Fig. 5. The linear and converging plots suggest that the data and the concept are thermodynamically sound.

The second reason for varying sugar sweetness recognised was that, as the glycol conformation deviated from the *gauche* arrangement toward the *anti* arrangement, sugar sweetness was diminished. For some reason, the vicinal OH groups were too far apart to cause sweet taste. This was resolved by the following reasoning: in order to

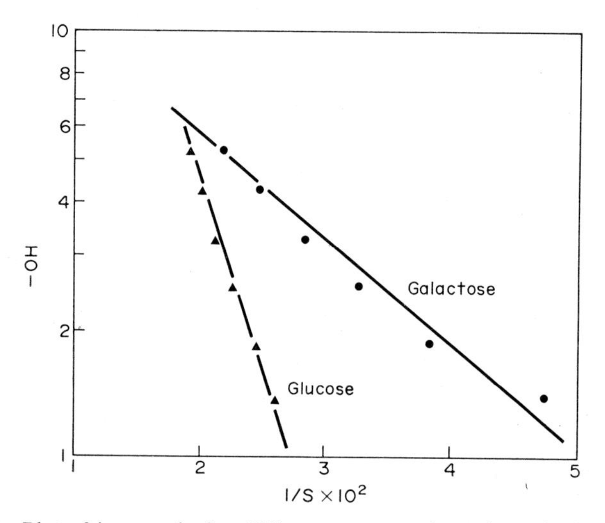

FIG. 5. Plot of increase in free OH groups *versus* the reciprocal of the sugar sweetness score.

cause varying sugar sweetness by the formation of intramolecular hydrogen bonds, it seemed probably that the initial chemistry of sweet taste was due to an intermolecular hydrogen bond between the glycol unit and a chemically commensurate group at the taste bud receptor site. Because we now had begun to think of varying sugar sweetness in terms of intramolecular H-bonding, and of the initial chemistry of the sweet taste response as due to intermolecular H-bonding, we began to think of the sugar glycol moiety as an AH,B system wherein A and B are electronegative atoms and H is a covalently bonded proton. This is the conventionally used system to define and describe the hydrogen bond. Thus, for the sugars, one glycol OH became AH, and the oxygen atom of the other became B. The AH proton to B orbital distance averaged out to be 3 Å for a *gauche* glycol moiety. It then followed that the receptor AH,B site must contain these same distance-parameters. Then hydrogen bonded glycol groups would be restricted in the intermolecular ability to bond the receptor AH,B. By the same token, *anti* sugar OH groups would be too far apart to bond intermolecularly to the receptor AH,B. This deduction is shown diagramatically in Fig. 6 and illustrates how we believed that the initial chemistry of the sweet taste response was due to a simultaneous interaction between AH,B

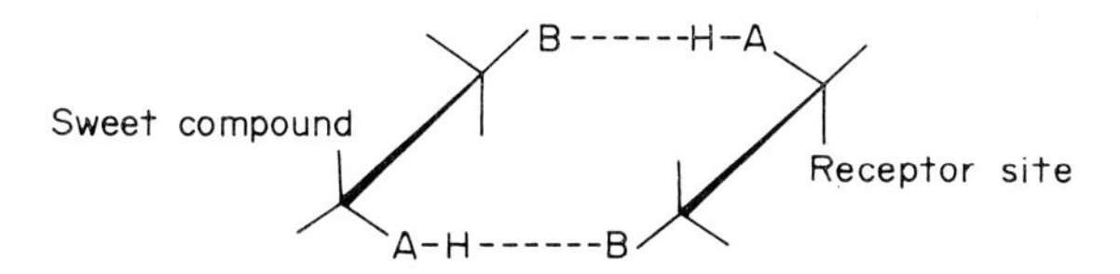

FIG. 6. Concerted interaction between the AH, B group of a sweet-tasting compound and AH, B of a taste bud receptor site.

of a sweet compound and AH,B of a receptor site. These chemical parameters seemed to us to have resolved the varying sweetness of the naturally occurring aldohexoses and D-fructose. Since all compounds which taste sweet possess an AH,B system with the appropriate AH proton to B distance-parameter, we proposed that this system is a prerequisite for any compound to taste sweet, and is the chemical element common to sweet compounds. Some of them are saccharin and the cyclamates, chloroform, and the D-amino acids.

THE STEREOCHEMICAL PROBLEM

At this stage of our studies, we encountered the stereochemical problem, not concerned so much with the sweet taste of diastereoisomers, but the sweet taste of enantiomers and its significance in relation to our chemical model for the receptor site.

When Emil Fischer was actively engaged in synthesising the amino acids, he tasted the preparations and found that the D-series generally tasted sweet, while the L-series did not. Speculation on this discovery led Louis Pasteur to state that the taste bud receptor site must, therefore, by asymmetric and mirror image compounds would give different responses. The model for the taste bud receptor site shown in Fig. 6 is symmetric and does not account for this established facet of taste.

After some thought, we elected to explain the difference between the D- and L-amino acids by erecting a spatial barrier behind the AH,B receptor site. Then an amino acid, with a fixed AH,B unit, could only make one approach to the site. Thus, D-leucine would taste sweet, as shown in Fig. 7, but L-leucine would not.

The sugar AH,B unit is the glycol group and since either OH substituent can act as either AH or B, the three-dimensional site

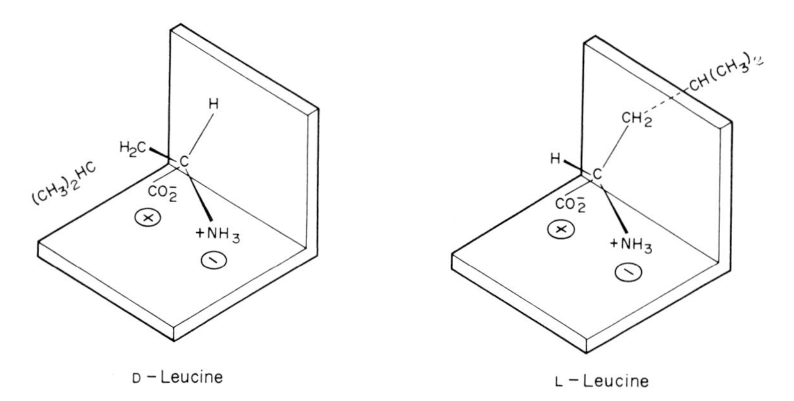

FIG. 7. Positioning of the sweet-tasting D-leucine over a receptor site, and the inability to position L-leucine over the site.

shown in Fig. 7 did not account for the supposed tastelessness of the L-forms. If the accounting of the varying sweetness of the amino acids was correct, then D- and L-sugars should be equally sweet. Upon obtaining various enantiomeric sugars, it was established that this was actually the case.

REFERENCES

1. Schallenberger, R. S. (1963). *J. Fd. Sci.*, **28,** p. 584.
2. Schallenberger, R. S. and Acree, T. E. (1967). *Nature, Lond.*, **216,** 4, p. 480.
3. Hawkins, R. I. and Schallenberger, R. S. (1970). *Nature, Lond.*, **227,** p. 965.
4. Cameron, A. T. (1947). *The taste sense and the relative sweetness of sugars and other sweet substances.* Sugar Res. Found., Inc., N.Y., Rept. No. 9.
5. Schallenberger, R. S. (1968). *Front. in Fd. Res.*, **June,** p. 40.
6. Brimacombe, J. S., Foster, A. B., Stacey, M. and Whiffen, D. H. (1958). *Tetrahedron*, **4,** p. 351.

DISCUSSION

Williams: Have any Raman studies been carried out on sugars (in general) and do such studies throw any light on relative sweetness, etc?
Shallenberger: Nothing that I know of.
Swain: It is interesting that β-D-mannose is bitter and that to aglycogeusia sufferers, fructose is acid (sour). Commonly, sweetness interacts with bitterness and acidity, and there is a balance between the taste pairs. Could you tell us whether sweet compounds interact with the bitter and acid taste buds to eliminate their responses to their respective stimuli?
Shallenberger: In answer to your last question, yes. Aglycogeusic people cannot taste sweetness. β-D-Mannose is a bitter compound and is unusual from configurational principles because its β-hydroxyl group bisects the bond-angle between the ring oxygen and the anomeric OH group. When this happens an element of conformational instability exists, and the molecule may flip from one form to the other.

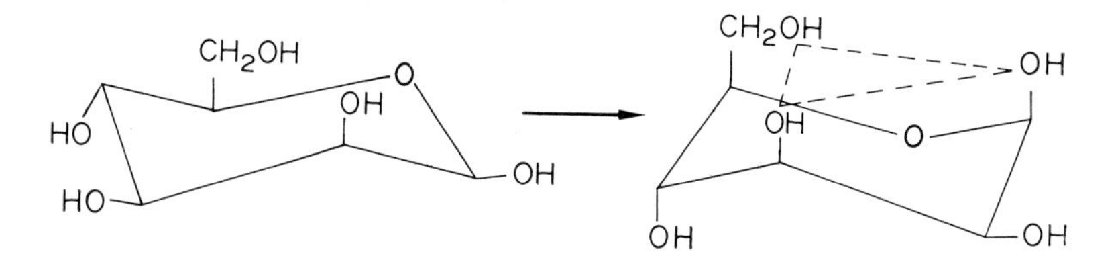

Whenever this happens three oxygen atoms now lie in the same plane. Intramolecular hydrogen bonding is associated with bitterness in some compounds.
Pangborn: α and β-D-Mannose configurations are both sweet and bitter, but this is a matter of degree only.
Noble: Kubota (1969), *Nature* (*Lond.*) **223,** p. 97, has suggested that the AH,B distance in some bitter compounds is 1·5 Å.
Shallenberger: I think that figure is far too low.
Spence: Professor Shallenberger has outlined how known sweet compounds fit into a pattern, explained by his theories. Would he care to comment on the many compounds with the conformational requirements, but which fail to exhibit any sweetness?
Shallenberger: Sweetness $= (\sigma)\ (\pi)$
where $\sigma =$ AH,B factor or electron-withdrawing power, and
$\pi =$ a hydrophobic factor governing access to receptor site
thus, the predictive power of the AH,B theory alone is nil.
Grenby: Could you give us information about the receptor sites in taste-buds that sweet molecules link on to?
Shallenberger: Well, of course, a protein can have a number of AH,B groups with the appropriate distance parameter. All this fits in with the concept of AH,B systems complexing by hydrogen-bonding to receptor sites. I cannot go beyond this.

Beidler: In reference to the difference of sweetness of sucrose and dextrose, is the strength of binding of the sugar to the receptor site related to the intensity of sweetness, or is there another factor?
Shallenberger: Yes, there is a relationship; sweetness is also related to the inflexibility of the AH,B site.
Fletcher: Many sweeteners appear to have an associated cool taste. Does Professor Shallenberger have any physical explanation along the lines of the sweetness correlation?
Shallenberger: β-D-glucose may have, in the IC conformation, a weak, ether-like linkage which causes a cooling effect.

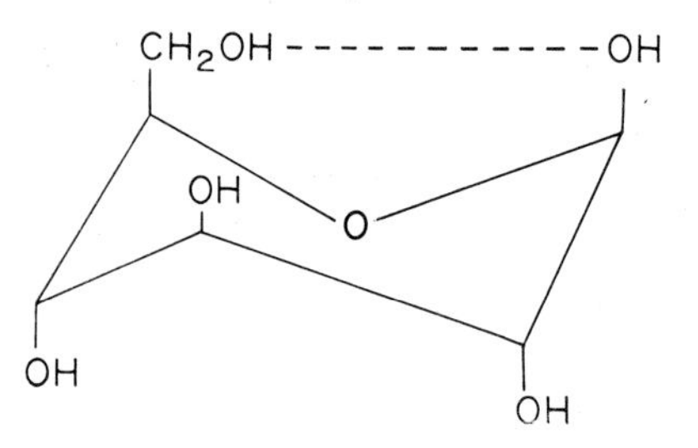

Arnold: Xylitol is both very sweet and very cool to the taste.
Birch: If β-D-fructopyranose is as sweet as Professor Shallenberger suggests, would not ethyl β-D-fructopyranoside, say, be the sweetest simple sugar known, and has Professor Shallenberger thought of making it?
Shallenberger: We have thought about a lot of things! One of the things we thought of in order to diminish calorific value is to synthesise such derivatives.

The Metabolism of Sodium Cyclamate

A. J. Collings

Environmental Safety Division,
Unilever Research Laboratory Colworth/Welwyn,
Colworth House, Sharnbrook, Bedford, England

ABSTRACT

It has been shown that about 25% of humans can convert (metabolise) to cyclohexylamine some of any ingested cyclamate. The absolute amount converted increases with an increase in cyclamate ingestion but in decreasing proportion. Conversion has been studied in man, rat and pig and it has been established that conversion is effected by microbial organisms in the lower intestinal tract. Cyclohexylamine blood levels found in human converters ingesting cyclamate have been compared with those in rats being fed cyclohexylamine. The results illustrate the difficulty in assessing the degree of risk in humans who can convert cyclamate to cyclohexylamine.

INTRODUCTION

Cyclamic acid (cyclohexane sulphamic acid) was first synthesised[1] in 1939. It was by accident that the salts were found to have sweetening properties and subsequent work showed that the calcium and sodium salts of cyclamic acid had approximately 40 times the sweetening power of sugar. Early toxicological investigations[2–5] showed that the cyclamate ion was only partially absorbed and that a proportion absorbed was excreted in the urine unchanged. Animal studies showed no untoward effect except for loosening of the stools at high dosage levels. This was later shown[6] to be due to the osmotic effect of the unabsorbed cyclamate on the large intestine. Cyclamates were permitted for use in the United States in about 1950 but toxicological investigations were continued. No striking results were obtained until 1967 when the detection of cyclohexylamine in the urine of a human after he had ingested sodium cyclamate was reported.[7] Previous to this there had been no reports of any metabolism or any metabolite of sodium cyclamate in any animal or man. Subsequent to the Japanese work it was found that in 45 subjects screened, 3 had the ability to convert some of the ingested cyclamate to cyclohexylamine[8] (Fig. 1).

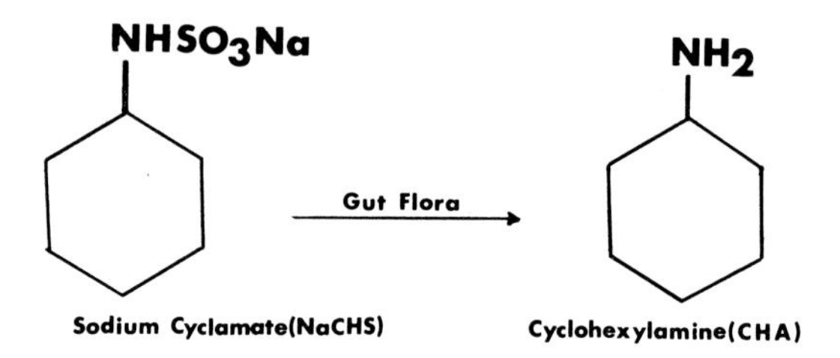

FIG. 1. Conversion of cyclamate to cyclohexylamine—the metabolism of sodium cyclamate in man.

Following this there was some confusion as to what really happened to cyclamate in man. Did conversion really occur or was cyclohexylamine normally present in the urine? It was at this time and in this confusion that Unilever started to take an active part in the study of cyclamate metabolism.

Our area of interest was restricted to metabolic studies and the work was designed to establish:

1. whether cyclohexylamine was present or not in normal human urine;
2. the incidence of converters in a normal population;
3. the relationship between the dose of cyclamate and the amount of cyclohexylamine produced;
4. the site of conversion; and
5. the relevance of the conversion of cyclamate in man.

INCIDENCE OF CONVERSION

One hundred of our staff were asked to avoid all sources of artificially sweetened food and drink for a week and then to provide an early morning urine sample. Using a gas–liquid chromatographic technique for cyclamate/cyclohexylamine analysis, cyclohexylamine was found in the urine of five subjects. These were placed on a really strict regime and subsequently four of the five stopped excreting cyclohexylamine. The one subject remaining, a young lady, continued to excrete cyclohexylamine in her urine. Subsequent investigation showed that she had been taking an iron tonic sweetened with cyclamate. This established that cyclohexylamine was not a normal metabolite in man but was associated with the metabolism of cyclamate.

200 g (1 g mole) of sodium cyclamate when converted produces 100 g (1 g mole) cyclohexylamine and this represents 100% conversion. In order to establish the incidence of converters among a normal population 3 surveys were carried out (Fig. 2). In each of

		Converters			
No. of subjects	Non-converters	0·15–1%	1–20%	20–60%	Over 60%
60	54	2	3	1	
43	35	2	5	1	
35	13	10	7	3	2
Children 3	3				
Total 141	105	14	15	5	2

FIG. 2. Incidence of converters—factors effecting conversion.

these the subjects were asked to take 0·5 g sodium cyclamate a day for four days and then to provide a urine sample. In the first survey 60 subjects were screened. Of these, 54 were non-converters, two converted small amounts, three converted in the 1–20% range and one in the 20–60% range. The second and third groups surveyed were taken from another laboratory and both screened within three weeks of each other. In the first screen, eight converters were found, whereas the second contained 22, including some who could convert very high proportions of the ingested dose. Three children have been screened, all three of whom were non-converters. The results show that about 25% of the people screened could convert cyclamate and some converted a very large proportion.

Effect of Cyclamate Dose on Conversion

Four volunteers who had the ability to convert cyclamate to cyclohexylamine were each given 250, 500 and 1000 mg of sodium cyclamate per day with two weeks on each dose level. Daily urine and faecal samples were collected for the last six days feeding at each dose level. As the dose of sodium cyclamate was increased the amount of cyclohexylamine being excreted also increased (Fig. 3). However, the percentage conversion decreased as the dose increased.

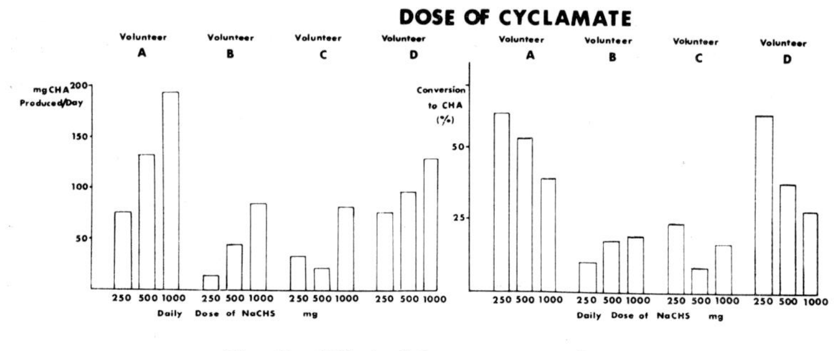

FIG. 3. Effect of dose on conversion.

This suggested that any hazard which might be associated with the production of cyclohexylamine from ingested cyclamate does not increase *pro rata* with the dose of cyclamate.

SITE OF CONVERSION

Absorption and Excretion Pattern

The fact that some humans have the ability to convert and others have not is an interesting problem in metabolism and in order to elucidate this, the excretion pattern of cyclamate and cyclohexylamine in a typical human converter was established. The subject was asked to avoid all cyclamate for at least a week and a urine spot test then showed the absence of both cyclamate and cyclohexylamine. The subject was then given half a gram of sodium cyclamate per day for 12 days. Twenty-four-hour urine samples were collected for cyclamate and cyclohexylamine analysis. The level of cyclamate being excreted in the urine in the first 24 hours was very high (Fig. 4) and subsequently extremely erratic. When the ingestion of cyclamate was stopped on day 12 the level of urinary cyclamate on day 13 was very small and was nil on day 14. Cyclohexylamine excretion took five days to reach an erratic maximum. This could be due to induction of enzymes, to the time taken for cyclamate to reach the conversion site or the accumulation of cyclohexylamine in the body.

At the end of the cyclamate ingestion the levels of urinary cyclohexylamine took at least a day before starting to decrease and then another two days before complete elimination occurred. This pattern is not typical of induction of enzymes but is consistent with an

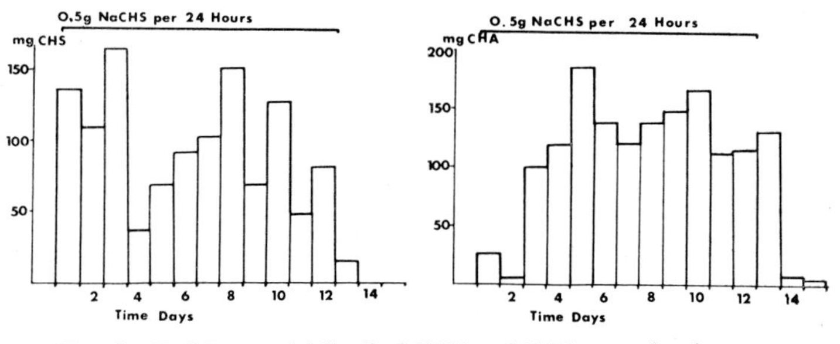

FIG. 4. Build-up and fall-off of CHS and CHA excretion in man.

accumulation of cyclohexylamine or the fact that the cyclamate takes time to reach the conversion site and is only slowly eliminated from it.

Effect of Antibiotic

A converter was given a daily dose of 750 mg of sodium cyclamate and his urine analysed for cyclohexylamine. He was then given a

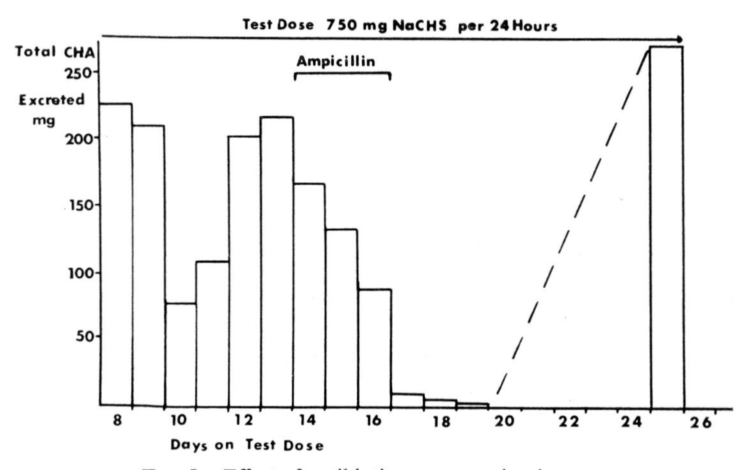

FIG. 5. Effect of antibiotic on conversion in man.

massive does of oral ampicillin, twice daily for three days. This reduced his gut flora to yeasts only. The effect on cyclohexylamine excretion was dramatic (Fig. 5). The pre-antibiotic excretion of cyclohexylamine was very erratic, but when the antibiotic was given, the level of conversion decreased almost to nil. However, when the gut flora were allowed to return to normal, a urine analysis showed that the subject had regained his ability to convert cyclamate to cyclohexylamine and was converting at the pre-antibiotic level.

CONVERSION IN ANIMALS

Pigs—Effect of Antibiotic

In the search for an experimental animal which could convert cyclamate to cyclohexylamine, it was found that our pigs were converters so they were used to extend the antibiotic study and to obtain additional information on the conversion site.

The animals were caged in our pig metabolism unit and given the sodium cyclamate in their diet. Daily urine and faeces were collected and analysed for both cyclamate and cyclohexylamine (Fig. 6).

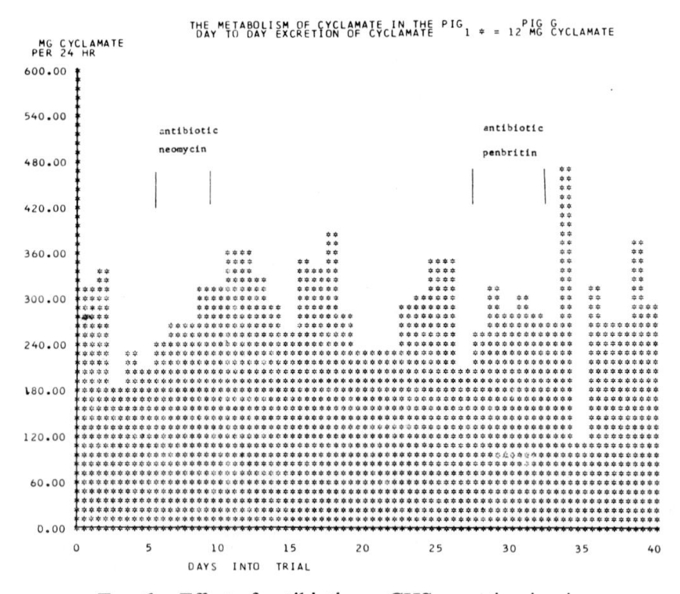

FIG. 6. Effect of antibiotic on CHS excretion in pig.

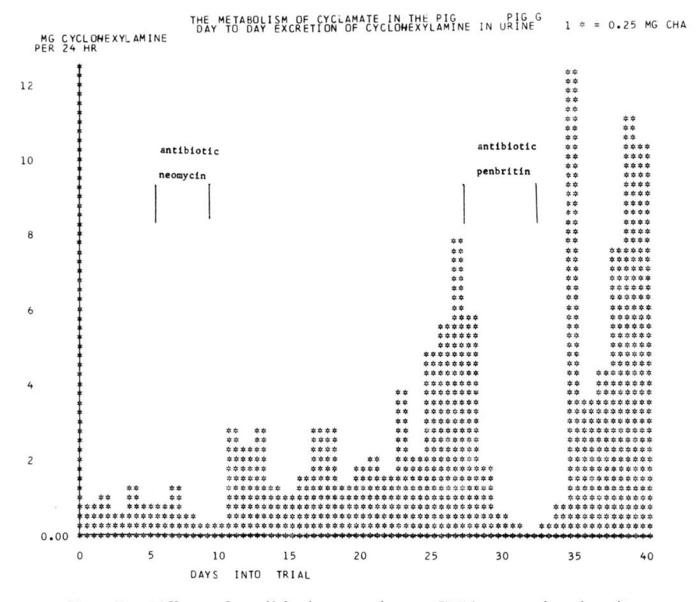

FIG. 7. Effect of antibiotic on urinary CHA excretion in pig.

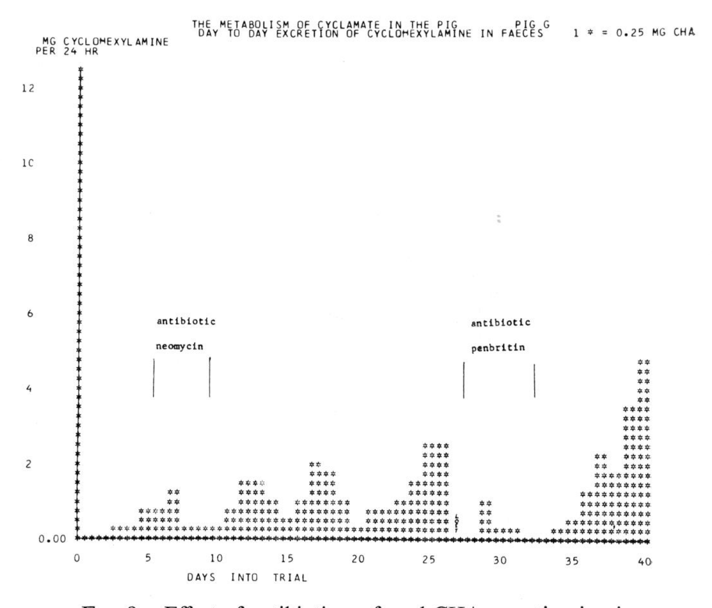

FIG. 8. Effect of antibiotic on faecal CHA excretion in pig.

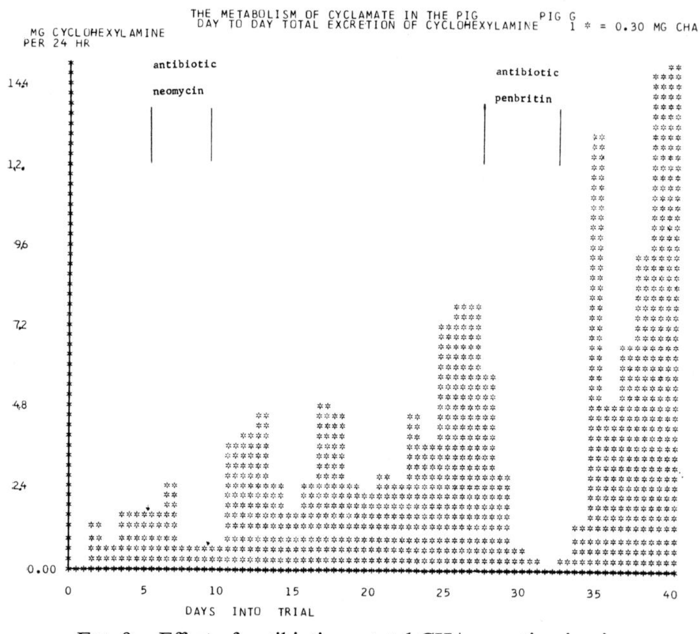

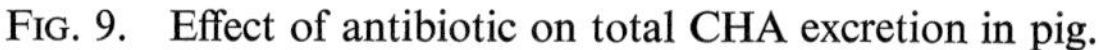
FIG. 9. Effect of antibiotic on total CHA excretion in pig.

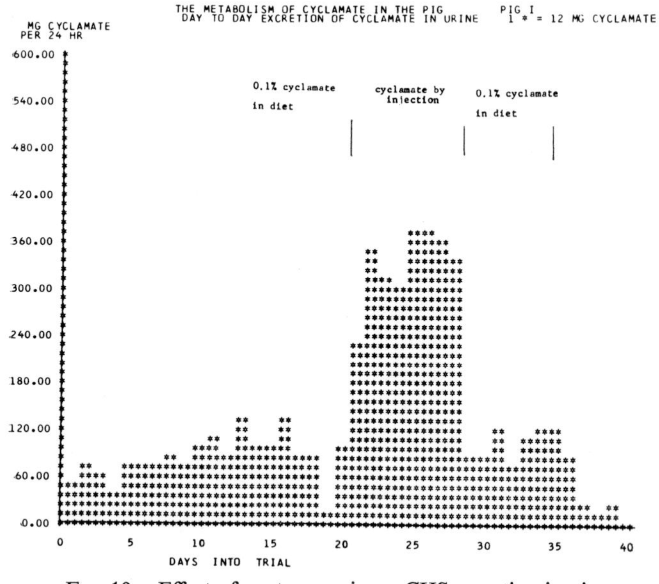

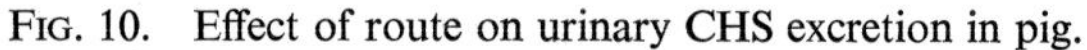
FIG. 10. Effect of route on urinary CHS excretion in pig.

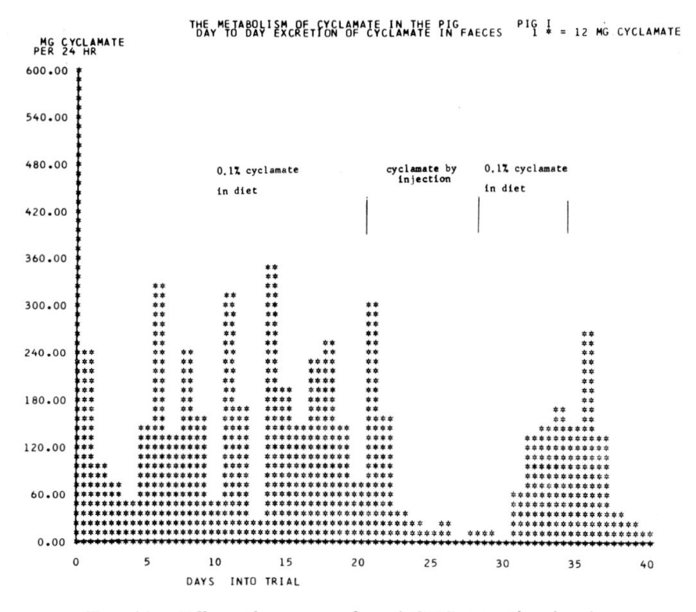

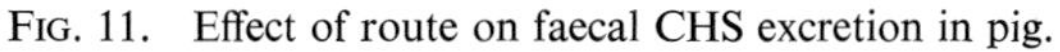
FIG. 11. Effect of route on faecal CHS excretion in pig.

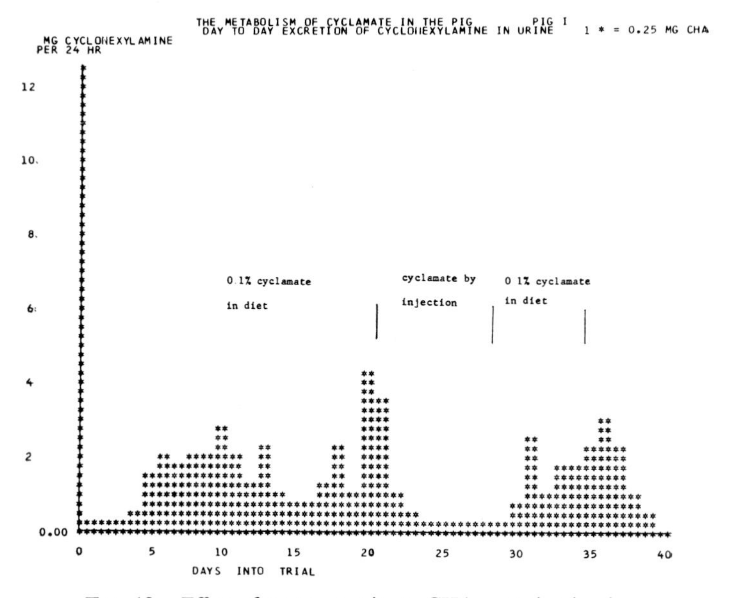

FIG. 12. Effect of route on urinary CHA excretion in pig.

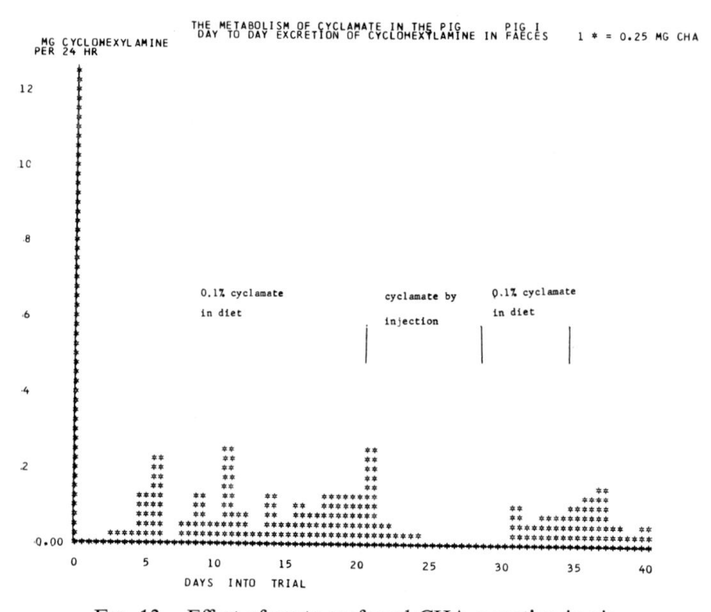

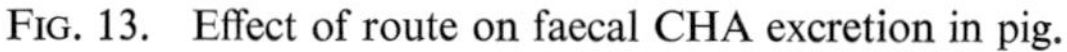

FIG. 13. Effect of route on faecal CHA excretion in pig.

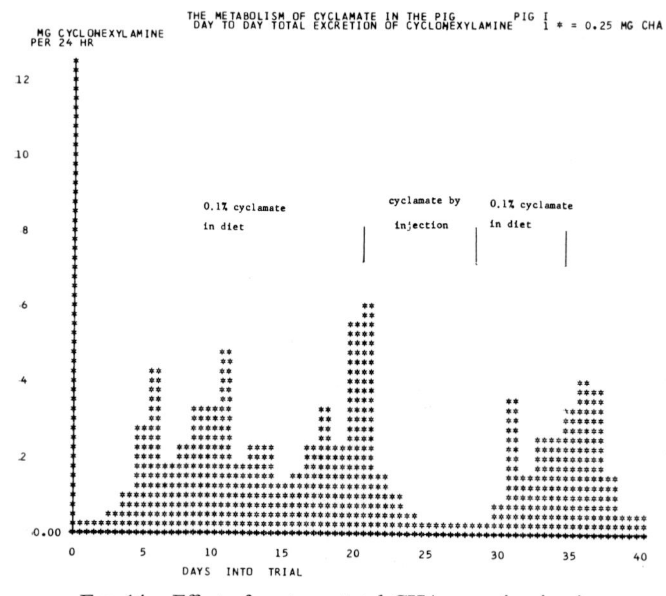

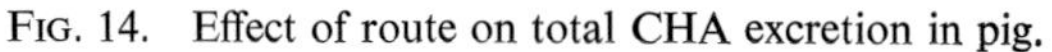

FIG. 14. Effect of route on total CHA excretion in pig.

With the Compliments of
Tate & Lyle Limited

Dr. K. J. Parker

Group Research & Development,
Philip Lyle Memorial Research Laboratory,
P.O. Box 68,
Reading, Berks., RG6 2BX.

Excretion of cyclamate was very erratic but typical of the excretion pattern for cyclamate. Oral antibiotic produced no detectable effect on cyclamate excretion. However, the urinary excretion of cyclohexylamine (Fig. 7), although erratic prior to the treatment, was significantly reduced. A similar effect was observed with the levels of faecal cyclohexylamine (Fig. 8). When the animals were taken off the antibiotic, the amount of cyclohexylamine excreted jumped very rapidly to levels higher than the pre-treatment levels (Fig. 9). The fact that both ampicillin and neomycin demonstrated this effect suggested that since neomycin is not absorbed it must be some micro-organism within the intestinal tract which is responsible for converting the cyclamate to cyclohexylamine.

Effect of Route on Conversion

A pair of pigs was kept on the cyclamate diet, and urine and faeces collected daily and analysed for cyclamate and cyclohexylamine. Cyclamate was then removed from the diet, but the same amount of cyclamate was given by subcutaneous injection (Fig. 10). The level of cyclamate being excreted in the urine rose immediately but fell equally rapidly when the animal was returned to the cyclamate diet and the injections stopped. Examination of faecal cyclamate excretion showed (Fig. 11) that, when the route was changed, excretion steadily decreased to zero. When the animal was returned to oral cyclamate, the faecal levels slowly built up again.

The pattern of excretion of cyclohexylamine in the pig was typically erratic (Fig. 12). When the cyclamate was given by injection, urinary cyclohexylamine steadily decreased to zero and when the animal was returned to the oral cyclamate, conversion gradually returned and cyclohexylamine excretion slowly built up to the pre-treatment level. Similarly faecal cyclohexylamine (Fig. 13) steadily fell to zero and gradually built up following a return to oral administration. The pattern of the fall-off of cyclohexylamine excretion (Fig. 14) was identical to that of faecal cyclamate and suggested that the presence of cyclamate in the intestinal tract is essential for cyclohexylamine production.

***In Vitro* Conversion**

At the end of this trial the animals were taken off the cyclamate regime and kept until the faeces were free of cyclamate and cyclohexylamine. The animals were slaughtered and the intestinal contents taken out and incubated with cyclamate under aerobic and anaerobic conditions. The contents of the large colon under anaerobic conditions, broke down some of the added cyclamate to cyclohexylamine, a clear demonstration that cyclamate is broken down by the gut flora to cyclohexylamine.

Transfer of Converting Ability

At Colworth House our rats, which are from a Caesarean derived colony but conventionally maintained, were fed on a diet containing sodium cyclamate for six months. Out of 100 animals on test not one acquired the ability to convert. However, we imported four rats from Holland which could convert at the 80–90% level. Our animals were put into contact with the imported animals and our animals acquired the ability to convert within three days, showing that conversion can be transferred from one animal to another. The fact that feeding cyclamate did not induce conversion suggested that induction of enzymes is not part of the process, but the fact that converting ability can be transferred from one animal to another is yet more evidence indicating that some micro-organism is involved.

Hazards Associated with Cyclohexylamine in Man

Cyclohexylamine is a stronger base than ammonia, and at physiological pH exists as the cyclohexylammonium ion. A dose of 3 mg/kg

MAN

CHA output mg/kg/day	CHA blood level ppm
1·46	0·3
2·5	1·0
3·56	0·2
0·13	0
0·54	0
1·13	0·2
0·48	0
0·23	0
1·24	0·1
1·5	0·1
1·9	0·4
2·8	0·4

RAT

CHA in Diet %	CHA intake mg/kg/day	CHA blood level ppm
1·0	475	11
0·5	191	2·8
0·20	75	0·45
0·1	37	0·2
0·05	19	0·02
0·01	3	0·02
Control	0	0

FIG. 15. CHA blood levels in human converters ingesting CHA and rats ingesting CHA.

of the hydrochloride by rapid intravenous injection causes a rise in blood pressure. The effect can be stopped by pre-emptying the adrenaline stores with reserpine. Cyclohexylamine (CHA) is very similar in its action to tyramine but is 20 times less active.[9]

When cyclohexylamine hydrochloride was fed to rats at the 0·1% dietary level and above, it induced toxic effects as demonstrated by their reduced body weight gain and a reduction in the relative size of their testes.[10] The 0·1% diet was equivalent to a daily intake of 37 mg CHA/kg and this produced a cyclohexylamine blood level of 0·2 ppm (Fig. 15). A similar blood level was found in human converters ingesting cyclamate and excreting 4 mg CHA/kg/day. In order to explain this disparity the absorption and excretion of cyclamate and cyclohexylamine were examined in rat.

Oral cyclamate is steadily absorbed and excreted in the urine, with conversion (Fig. 16), whereas cyclamate by subcutaneous injection is very rapidly excreted without conversion (Fig. 17). From a pharmacodynamic point of view cyclamate is an extremely desirable food additive; it is slowly absorbed and the proportion that is absorbed is rapidly excreted.

Oral cyclohexylamine hydrochloride is completely excreted in the urine within 16 hours. The blood cyclohexylamine levels following oral administration are typical of a substance which is very rapidly absorbed and excreted (Figs. 18 and 19).

It could be speculated that in the rat feeding study, cyclohexylamine wastaken in through the diet at discreet times, mainly during

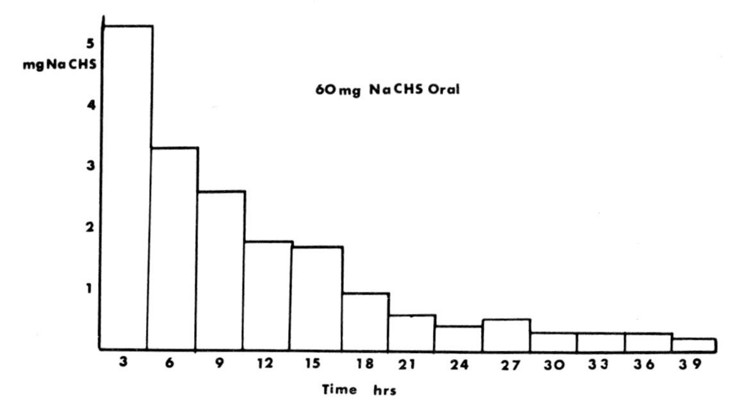

FIG. 16. Absorption and excretion of oral CHS in rat.

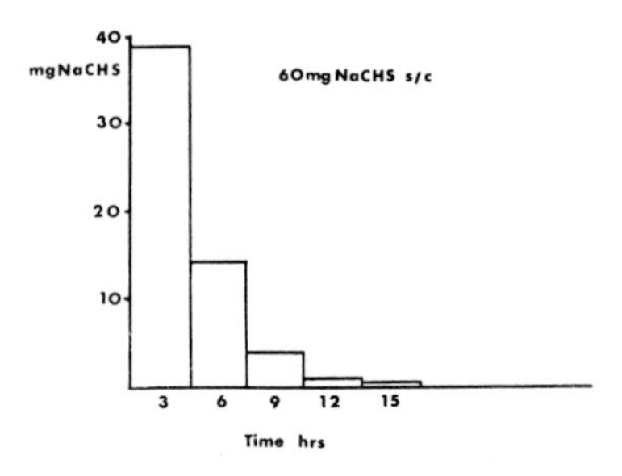

FIG. 17. Excretion of CHS in rat following subcutaneous administration.

the night. Since cyclohexylamine is so rapidly absorbed and excreted, the blood levels in the rats would be expected to vary from hour to hour with peak levels occurring after the ingestion of food. The blood samples were taken some three hours after the last food had been ingested and by this time the blood levels would be well past their maximum.

In man, the cyclohexylamine would probably be produced throughout the day by the action of the gut flora on the unabsorbed cyclamate as it passed through the intestinal tract. Once formed, the cyclohexylamine would rapidly be absorbed and excreted and the

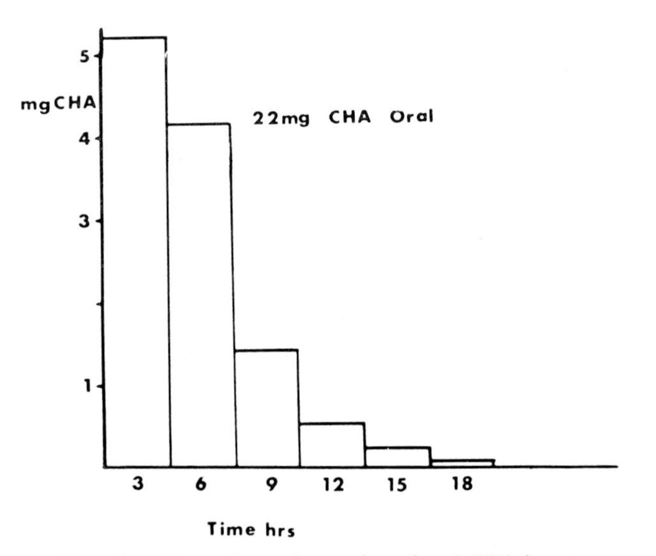

FIG. 18. Absorption and excretion of oral CHA in rat.

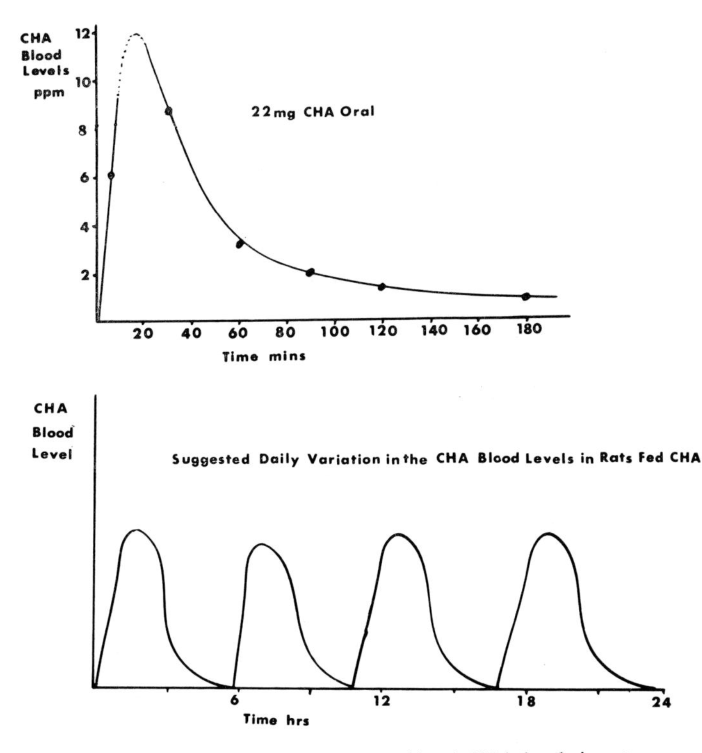

FIG. 19. Effect of oral CHA on blood CHA levels in rat.

blood would, therefore, be in the steady state between production and excretion of cyclohexylamine.

This apparent disparity between the human and rat cyclohexylamine blood levels could therefore be due to the variation in the rat levels caused by taking cyclohexylamine through the diet.

However, this serves as an example of the great difficulty in estimating the degree of risk associated with a metabolite produced from a food additive by only certain humans. Feeding the metabolite in the diet induces toxic effects but are these effects caused by the transient high blood levels or the much longer exposure to low levels? Quite clearly, if cyclamates are ever reconsidered as food additives, these factors must be taken into full consideration.

In concluding, I should like to say that this work was initiated well before the ban on cyclamates following the reports of induced bladder cancer[11] but the blood level work was completed after the ban. All these data were made available, at the earliest opportunity, to the relevant Government Committees in the United Kingdom and to the Joint FAO/WHO Expert Committee on Food Additives.

REFERENCES

1. Audrieth, L. F. and Sveda, M. J. (1944). *J. Org. Chem.*, **9,** p. 89.
2. Richards, P. K., Taylor, J. D., O'Brien, J. L. and Deusher, H. O. (1951). *J. Am. pharm. Ass.*, **40,** p. 1.
3. Schoenberger, J. A., Rix, D. M., Sakamoto, A., Taylor, J. D. and Kark, R. M. (1953). *Am. J. Med. Sci.*, **5,** p. 225.
4. Fitzhugh, O. G. and Nelson, A. A. (1950). *Fedl. Proc.*, **9,** p. 272.
5. Taylor, J. D., Richards, R. K. and Davin, J. C. (1951). *Proc. Soc. Exp. Biol. & Med.*, **78,** p. 530.
6. Hwang, K (1966). *Arch. int. Pharmacodyn. Ther.*, **163,** p. 302.
7. Kojima, S. and Ichibagase, H. (1966). *Chem. Bull. Tokyo*, **14,** p. 971.
8. Leahy, J. S., Wakefield, M. and Taylor, T. (1967). *Fd. Cosmet. Toxicol.*, **5,** p. 447.
9. Cambridge, G. W. and Parsons, J. (1970). *Unilever Research Laboratory, Colworth House*, unpublished results.
10. Collings, A. J. and Kirkby, W. W. (1970). *Unilever Research Laboratory, Colworth House*, unpublished results.
11. Price, J. M., Biava, C. G., Oser, B. L., Vogin, E. E., Steinfeld, J. and Ley, H. L. (1970). *Science, New York*, **167,** p. 1132.

DISCUSSION

Renwick: There has been a suggestion that cyclohexylamine only represents 'a tip of an iceberg' of total metabolism. Have you any information on the extent of cyclohexylamine metabolism?
Collings: The analytical techniques we have used will detect cyclamic acid, cyclohexanol, cyclohexene and cyclohexylamine. We have only found cyclohexylamine in our samples; no trace of other metabolites was found.
Walker: 1. Was all the work done with sodium cyclamate; has Dr Collings any evidence obtained with the calcium salt?

2. Relating to the first question, what overall recoveries were observed as cyclamic acid, cyclohexylamine and metabolites?

3. If the ability to convert cyclamic acid to cyclohexylamine resides in the gut microflora, ought we not to consider all humans as *potential* 'converters'?
Collings: 1. We have worked only with sodium cyclamate on both humans and rats.

2. In the human sets of experiments, recoveries in general were 90–95%, and animal recoveries have been in the same region.

3. The problems concerning gut flora are extremely complex and no relationship between 'non-converters' and 'converters' has been established.

Elias: 1. Did you establish a 'no effect level' in the 90-day test on cyclohexylamine in rats?

2. Have you done any long-term studies on cyclohexylamine?

Collings: 1. The no-effect level on the 90-day test was 0.05% of the dietary level.

2. The longest we fed cyclohexylamine was 90 days.

Wiggall: The speaker mentioned that the ability to convert cyclamic acid to cyclohexylamine was transferable in the rat and other animals. He also mentioned that in the laboratory personnel examined, the ability to 'convert' appeared to be associated in groups. Does Dr Collings consider that we are all, in fact, potential 'converters', in view of these two observations?

Collings: We have looked at one family and only Dad 'converted'. We are not all potential 'converters'; not all animals 'convert', either. As far as humans are concerned, once a 'converter', always a 'converter'; and *vice versa*.

Grenby: What exactly are the toxic effects of cyclohexylamine in the body?

Collings: Atrophy of the testes and associated effects.

Galitzine: Is it true that there were important differences in the make-up of the Japanese-made cyclamic acid which meant that it produced a higher incidence of cyclohexylamine?

Collings: There was no difference in the proportion of material 'converted' between pure, *e.g.* CIBA-produced cyclamic acid and Japanese-produced products containing up to 100 ppm cyclohexylamine.

Walker: Do all potential 'converters' have appropriate gut flora?

Collings: There seems to be no evidence of this.

Gurnley: Have you found any evidence of conversion to dicyclohexylamine?

Collings: We have found no evidence of this. We did in fact have somewhat of a scare when we picked up an agent in the urine in some of our volunteers, but we found it was nicotine!

Paine: If cyclamic acid were to be reconsidered, at what point would the testing programme start?

Collings: To survey presently available information; assess the toxicology of cyclohexylamine and then long-term feeding studies.

Wood: Does the rate of conversion of *pure* cyclamic acid differ from the rate when taken as part of a diet?

Collings: There is no difference.

Noble: Could you say a few words on cyclamate and its relationship to bladder cancer? The massive doses of cyclamate given to rats invalidate the results, don't they?

Collings: In a study which was carried out by the Abbott Laboratory involving cyclamate, saccharin and in the later part of the trial, cyclohexylamine as well, the evidence produced was indicative of bladder cancer, and under such circumstances when a substance is thought to be potentially cancerous, there is no choice but to withdraw it until such time as more evidence is available. Since the Abbott work, other two-year studies on cyclamates have been completed, but I am not aware that the results have been officially published. The dose administered was extremely large. The experiment was designed to show that cyclamate/saccharin mixtures were *safe*. If they had had no effect at all, then we should have been perfectly happy, but the results threw everyone into complete turmoil. This raises the question of the quality of the original experiment, but it was obviously right to err on the safe side.

Renwick: There is a report that 3 out of 23 rats given oral doses of sodium cyclamate (400 mg/kg/day) for prolonged periods developed tumours of the urinary bladder.

Dihydrochalcone Sweeteners

R. M. Horowitz and Bruno Gentili

United States Department of Agriculture,
Agriculture Research Service,
Fruit and Vegetable Chemistry Laboratory,
Pasadena, California, USA

ABSTRACT

Certain intensely bitter flavanone glycosides occur in citrus fruits together with related glycosides that are tasteless. The bitter compounds all contain the disaccharide β-neohesperidose (2-O-*α*-L-*rhamnosyl-β*-D-*glucose*). *The tasteless compounds contain an isomeric disaccharide, β-rutinose* (6-O-*α*-L-*rhamnosyl-β*-D-*glucose*). *Isomerism in the sugar moiety is thus of crucial importance in determining bitterness or tastelessness in this group of compounds. When alterations are made at selected sites in the flavanone neohesperidosides, the product may be bitter, bitter-sweet, sweet or tasteless. Corresponding changes made in the flavanone rutinosides usually result in tasteless compounds unless the rhamnose is removed from the* 6-*position of glucose.*

Of particular significance in this series are the dihydrochalcone neohesperidosides and glucosides prepared by catalytic hydrogenation of the chalcone form of the flavanones. Several of these compounds are intensely sweet. From a practical point of view the most interesting of these sweeteners are naringin dihydrochalcone (derived from naringin), neohesperidin dihydrochalcone (derived from neohesperidin or the conversion of naringin) and hesperitin dihydrochalcone glucoside (derived from hesperidin). Many variations of the compounds have been made in order to clarify structure-activity relations. The conclusions that can be drawn from these studies will be summarised and a review of earlier work given.

NARINGIN

In 1958 we undertook to study the relation between taste and structure of the phenolic glycosides occurring in citrus. We were concerned chiefly with naringin, a well-known bitter principle of grapefruit. Earlier work had shown that naringin is the flavanone rhamnosyl-β-glucoside **1**, in which the point of attachment of rhamnose to glucose is left unspecified. The structure of a related citrus glycoside, hesperidin, was known in almost complete detail. It is the 7-β-rutinoside of the flavanone aglycone 2(S)-hesperetin and has structure **2**. A perplexing feature was the fact that naringin

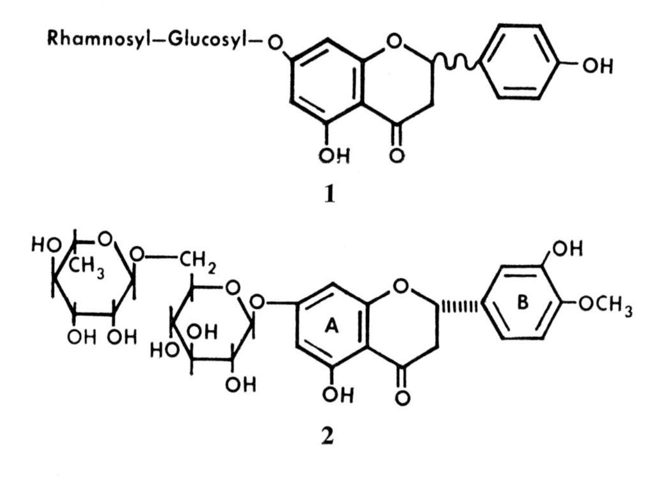

had been described in the literature as a rutinoside differing from hesperidin only in the substitution pattern of the B-ring, yet hesperidin itself is tasteless. We assumed it was unlikely that what appeared to be rather minor structural differences in the B-ring could account for the marked difference in taste. Today—many years and many compounds later—we know that this was a shaky assumption. Nevertheless, it did lead us into a most fascinating area of research.

At the outset of our experiments we methylated naringin completely, so that all the phenolic and alcoholic hydroxyl groups were affected. Acid hydrolysis of the product yielded two methylated sugars, which were shown to be 2,3,4-tri-*O*-methyl-L-rhamnose (**3**) and 3,4,6-tri-*O*-methyl-D-glucose (**4**). The identity of these sugars,

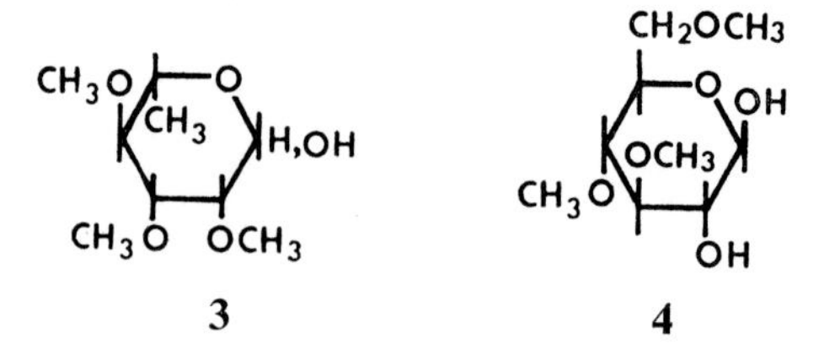

together with data on optical rotations, proved that naringin must be **5**, in which the disaccharide component is 2-*O*-α-L-rhamnopyranosyl-β-D-glucopyranose. At the same time we showed that a

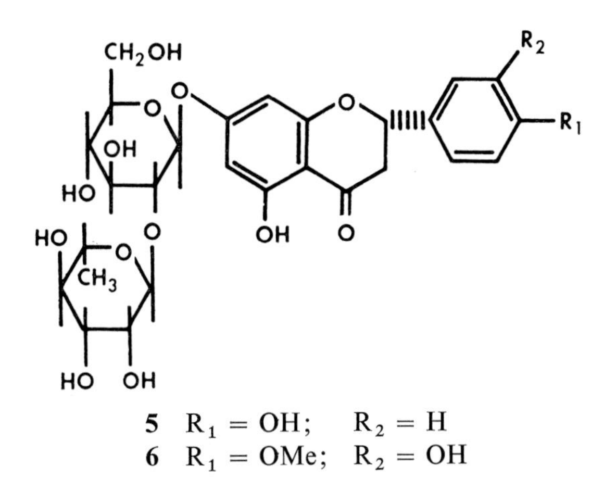

5 R_1 = OH; R_2 = H
6 R_1 = OMe; R_2 = OH

third citrus flavanone, neohesperidin, contains the same disaccharide and can be written as **6**. This disaccharide, named before its structure was known, is called neohesperidose.

The results implied that the bitterness of naringin *versus* the tastelessness of hesperidin is due to the difference in the point of

TABLE 1
Flavanone glycosides of citrus fruits

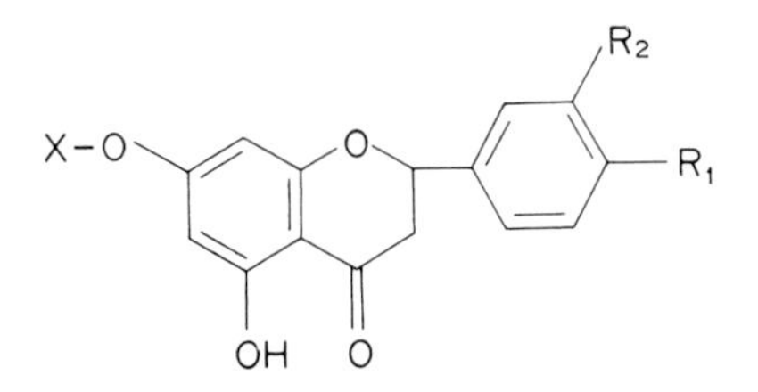

X = Rutinosyl	X = Neohesperidosyl	R_1	R_2
Hesperidin	Neohesperidin	OMe	OH
Naringenin rutinoside	Naringin	OH	H
Isosakuranetin rutinoside	Poncirin	OMe	H
Eriocitrin	Neoeriocitrin	OH	OH

attachment of rhamnose to glucose. This belief was reinforced when we found that neohesperidin is bitter. Here the only difference between the bitter neohesperidin and tasteless hesperidin lies in the point of attachment of rhamnose to glucose. As a result of further isolation studies we now know of the existence of four flavanones, each of which is represented, in various citrus species or hybrids, by both a rutinosyl and neohesperidosyl derivative (Table 1). A significant feature is that every flavanone 7-*β*-neohesperidoside shown in Table 1 is bitter, while every flavanone 7-*β*-rutinoside is tasteless.

STRUCTURAL REQUIREMENTS FOR TASTE

Having established the relationship between disaccharide structure and bitterness for this group of compounds, we then endeavoured to map out some of the other structural requirements for taste. We found, for example, that the presence of rhamnose at the 2-position of glucose is not mandatory for bitterness (**7** and **8** are bitter), though in some instances it may enhance the taste (bitter or sweet, as the case may be) by an order of magnitude or more. In contrast, rhamnose at the 6-position of glucose seems invariably to abolish taste responses, regardless of the nature of the aglycone. That the B-ring is not required for bitterness is illustrated by compound **9**, which is bitter.

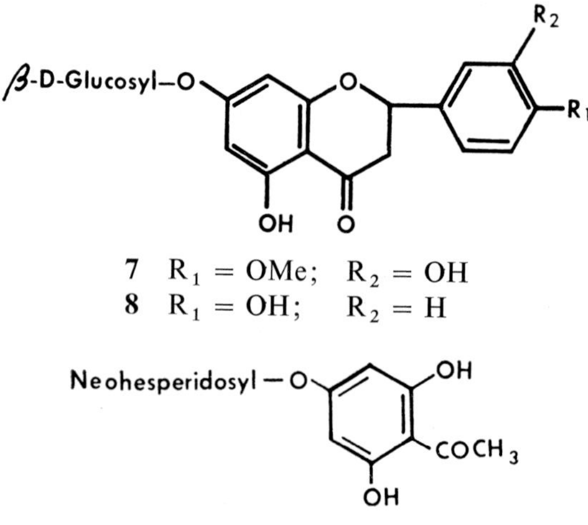

One of the transformations we tried was the conversion of naringin **5** to the chalcone **10** followed by catalytic hydrogenation to the dihydrochalcone **11** (here Neo stands for *β*-neohesperidosyl) (eqn. 1). We anticipated, for reasons that are no longer pertinent, that **10**

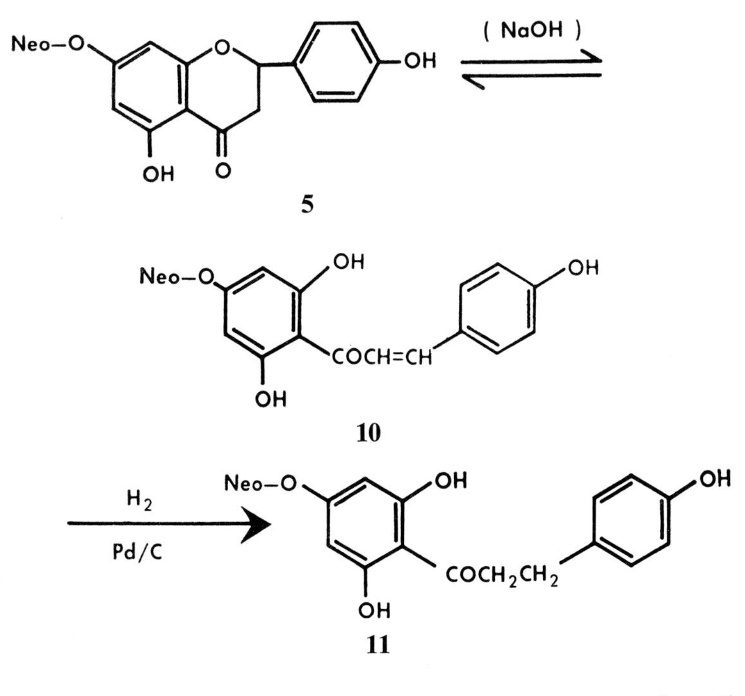

(eqn. 1)

would be tasteless and **11** bitter. Surprisingly, both **10** and **11** were intensely sweet. This result led us to study the conversion of the remaining three flavanones in Table 1 to see whether their dihydrochalcones might be sweet. Only one of these, neohesperidin, yielded a sweet dihydrochalcone (Table 2). The dihydrochalcone from poncirin was mainly bitter and that from neoeriocitrin was, at most, slightly sweet.

At present there are three dihydrochalcones which claim our attention as potentially useful sweeteners. These are naringin dihydrochalcone (**11**), neohesperidin dihydrochalcone (**12**) and hesperetin dihydrochalcone 4′-*β*-D-glucose (**13**). As mentioned,

TABLE 2

Taste and relative sweetness of dihydrochalcones and saccharin

Compound	Taste	Relative sweetness Molar	Relative sweetness Weight
Naringin dihydrochalcone	Sweet	1	0·4
Neohesperidin dihydrochalcone	Sweet	20	7
Poncirin dihydrochalcone	Slightly bitter	—	—
Neoeriocitrin dihydrochalcone	Slightly sweet	—	—
Hesperetin dihydrochalcone glucoside	Sweet	1	0·4
Saccharin (Na)	Sweet	1	1

11 is prepared from naringin, a compound easily available from grapefruit. **12** is prepared from neohesperidin, one of the major flavonoid constituents of Seville oranges (*C. aurantium*). Since there is relatively little of this fruit grown, neohesperidin is scarce and

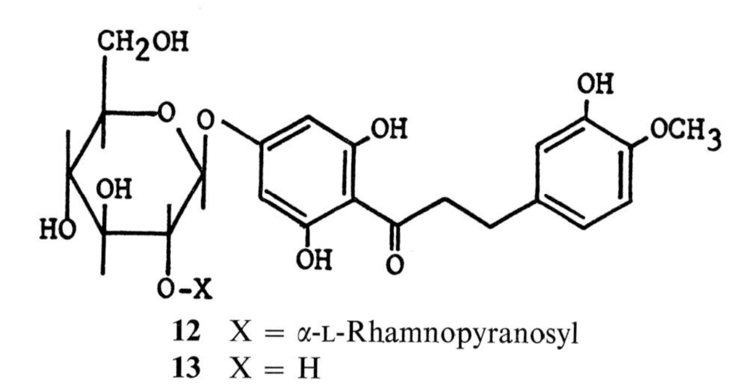

12 X = α-L-Rhamnopyranosyl
13 X = H

it is usually necessary to prepare **12** by the conversion of naringin (eqn. 2). **13** is most conveniently obtained by converting

$$\mathbf{5} \xrightarrow{\text{NaOH}} \mathbf{9} \xrightarrow{\text{Isovanillin}} \xrightarrow[\text{Pd/C}]{H_2} \mathbf{12} \qquad \text{(eqn. 2)}$$

hesperidin **2** to its dihydrochalcone (a tasteless compound), followed by partial hydrolysis to remove rhamnose. Hesperidin is abundantly available from sweet oranges (*C. sinensis*) and lemons (*C. limon*).

Of the three compounds neohesperidin dihydrochalcone is outstanding for its high level of sweetness and solubility. Although each compound has somewhat different taste characteristics, in general they exhibit a pleasant sweetness, rather slow in its onset and of varying (usually long) duration. There is no bitter aftertaste, but there is a sensation vaguely reminiscent of licorice or menthol.

STRUCTURE-ACTIVITY RELATIONS

We turn to a more detailed consideration of the structure-activity relations in this series. Since 1961, when the first sweet dihydrochalcone was prepared, many variations have been tried both in our own laboratory and elsewhere. Most of the variations have centered upon the hydroxyl and alkoxyl groups of the B-ring because it is relatively easy to make alterations in this part of the molecule. Depending on the substitution pattern the compounds can be divided into four groups. In the following discussion R in the structural formulae stands for β-neohesperidosyl-O-Ar-$COCH_2CH_2$-; + signifies sweet; − signifies bitter; and −+ signifies bittersweet.

1. Hydroxy Substituted Dihydrochalcones

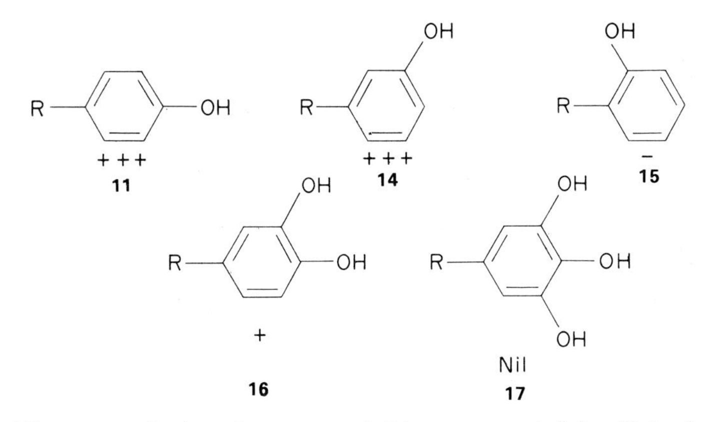

The *meta*-substituted compound **14** was reported by Krbechek *et al.*[1] to be as sweet as naringin dihydrochalcone **11**, while the *ortho*-compound **15** was said to be bitter. The disubstituted neoeriocitrin dihydrochalcone **16** is only slightly sweet and the pyrogallol derivative **17** is tasteless.

2. Hydroxy-Alkoxy Disubstituted Dihydrochalcones

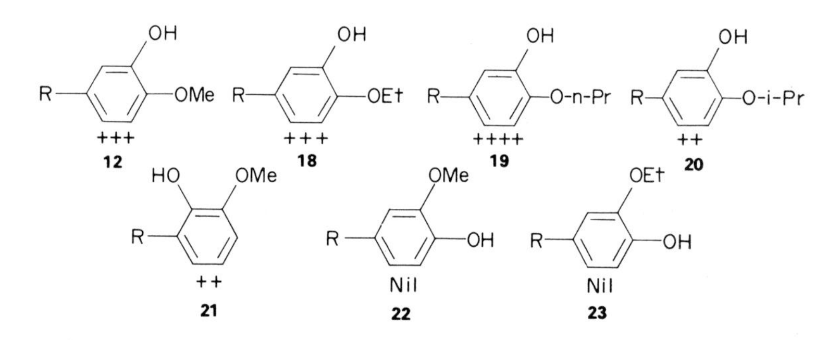

The sweetest compounds of the series belong to this group. Krbechek *et al.*[1] found the ethyl derivative **18** to be about as sweet as neohesperidin dihydrochalcone **12**, but the n-propyl derivative **19** was twice as sweet. The isopropyl derivative **20** is less sweet than any of the preceding three. From a practical point of view, a drawback to the dihydrochalcones with lengthened side chains is the difficulty in preparing the properly substituted aldehydes needed for their synthesis.

We found compound **21** to be moderately sweet. Quite unexpected, however, was the effect of reversing the hydroxy and alkoxy substituents of **12** and **18** to give **22** and **23**. Both were tasteless.

3. Hydroxy-Alkoxy Trisubstituted Dihydrochalcones

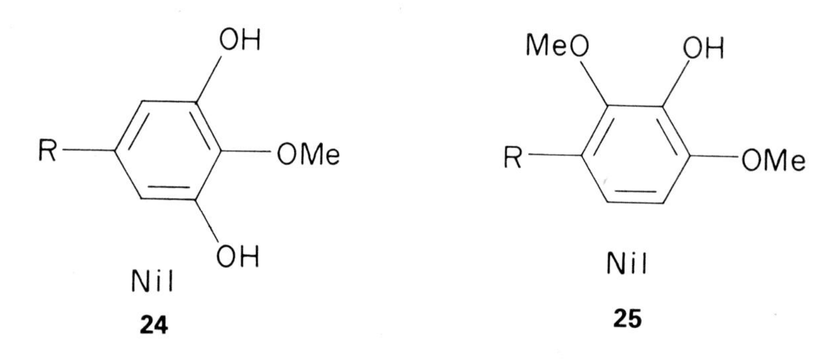

Before embarking on the troublesome task of preparing **24** we were confident it would be sweet because of the relation it bears to **12**. In fact, neither **24** nor **25** exhibits any sweetness whatsoever.

4. Dihydrochalcones with no Hydroxy Substituents

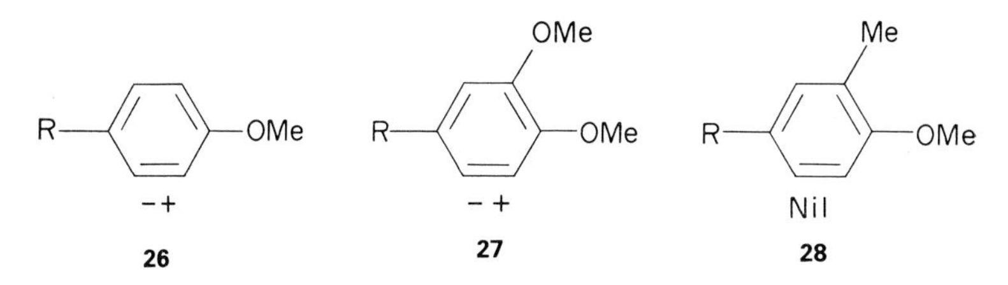

Poncirin dihydrochalcone **26** and the methoxylated analogue **27** are mostly bitter, though they do have a trace of sweetness. The *C*-methyl derivative **28** is tasteless.

The data for the four groups of compounds allow several conclusions to be made about structure-activity relations in this series.

(1) A hydroxyl group is necessary for sweetness (**11**, **14**, **16**, **12**, **18**, **19**, **20**, **21**) but its presence does not guarantee it (**15**, **17**, **22**, **23**, **24**, **25**).

(2) The absence of a hydroxyl group assures non-sweetness (**28**) or bitter-sweetness (**26**, **27**).

(3) For sweetness to subsist in hydroxy-alkoxy disubstituted compounds the order of the groups must be R-H-OH-Alkoxy (**12**, **18**, **19**, **20**) or R-OH-Alkoxy (**21**).

(4) Taste is abolished if the order of groups in hydroxy-alkoxy disubstituted compounds is R-H-Alkoxy-OH (**22**, **23**). It is also abolished if three adjacent groups are present in addition to R (**17**, **24**, **25**).

Dihydrochalcone-Receptor Complexes

To a certain extent these results can be rationalised in terms of a hypothetical taste receptor. Complexing between a receptor and the sapid compound is believed to be the initial event leading to the taste response. The postulated dihydrochalcone-receptor complexes are represented schematically in Figs. 1–4. We assume that, as a minimum, the receptor has sites that respond to (1) hydroxyl, (2) alkoxyl and (3) the glycosidic and adjacent parts of the molecule (R in the previous structural formulae). We are concerned here with the part of the receptor that complexes with the hydroxy and alkoxy substituents on the B-ring. We assume that this part of the receptor has an arrangement of surfaces which allow free entry of the molecule into the receptor but which limit the possible twisting and rotational orientations of the B-ring. In Figs. 1–4 the site for R is represented as a rectangular area; that for hydroxy as a triangular area; and that for alkoxy as a circular area.

Figure 1 represents the 'best' complex, *i.e.* the one for the sweetest compound, neohesperidin dihydrochalcone, **12**. Figure 2 is the complex for another sweet compound, **14**. Here we expect complexing even in the absence of the alkoxyl group but, since there is

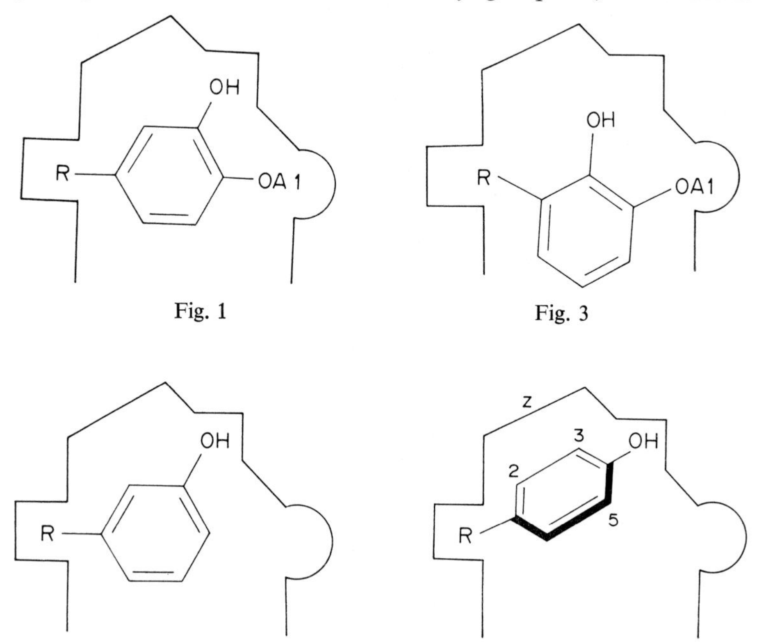

Fig. 1 Fig. 3

Fig. 2 Fig. 4

diminished binding to the receptor, it is reasonable that the sweetness of **14** is only about 1/20 that of **12**. Figure 3 represents the complex with another sweetener, **21**. We assume it is permissible to twist the B-ring clockwise to provide the fit, since the large R group should be flexible. Figure 4 represents the complex with naringin dihydrochalcone, **11**. To obtain the proper fit we have, first, twisted the B-ring counterclockwise and, secondly, rotated the ring out of the vertical plane along the R—OH axis in order to minimise interference between the upper surface, designated z in Fig. 4 and the protons at the 2 and 3 positions.

Figures 1–4 show how the four sweet dihydrochalcones could

complex with the hypothetical receptor. What can we say about the non-sweet or bitter-sweet ones? If we add an alkoxyl group at position 5 of the compound in Fig. 4 we obtain the non-sweet compounds **22** and **23**. Because of the partial rotation of the ring on its axis the alkoxyl would be displaced from the site it would normally occupy on the receptor. Moreover, since the alkoxyl group is much bulkier than the proton it replaced, interaction of this group with the surrounding vertical surfaces would substantially reduce the partial rotation of the B-ring. This, in turn, would make it unlikely that effective hydroxy-receptor complexing of the type shown in Fig. 4 could occur. Similar reasoning would apply to the slightly sweet dihydroxy compound **16** and the trisubstituted derivatives **17**, **24** and **25**. We can visualise weak complexing with the bitter-sweet substances **26** and **27**, but here, as with the *ortho*-hydroxy derivative **15**, the picture is complicated by competition for the compounds from other receptors, *i.e.* those involved in bitterness. However rudimentary and inexact these views on receptor topology may be, we believe they can provide guidelines and have predictive value for future studies in this field

REFERENCES

1. Krbechek, L., Inglett, G., Holik, M., Dowling, B., Wagner, R. and Riter, R. (1968). *Agric. and Fd. Chem.*, **16,** p. 108.
2. Horowitz, R. M. and Gentili, B. (1963). *US Patent* 3,087,821 (April 30, 1963).
3. Horowitz, R. M. (1964). In *Biochemistry of Phenolic Compounds*, Chapter 14, Ed. by Harborne, J. B., pp. 545–71, Academic Press, New York.
4. Horowitz, R. M., Gentili, B., Booth, A. N. and Robbins, D. J. (1968). *Agricultural Research Service Publication CA* 74-18, Western Regional Research Laboratory, US Department of Agriculture, Albany, California 94710.
5. Horowitz, R. M. and Gentili, B. (1969). *Agric. and Fd. Chem.*, **17,** pp. 696–700.

DISCUSSION

Charley: In view of the accepted non-toxicity of flavonoids themselves, what are the possibilities of the hydrogenated chalcones having toxic effects?

Horowitz: As you point out, flavonoids as a group are notoriously innocuous. Dr Booth has tested dihydrochalcones for as long as 170 days on rats, giving them a dosage as high as 5% of the diet, and, at this level, there were no effects. At the present time, he is on a two-year study on dogs and rats. The rat study has been under way, now, for about six months, but we need about 400 lb before we begin on dogs. A great deal is known about flavonoid chemistry from the work of Booth, and their safety is associated with the fact that they do not contain nitrogen or sulphur atoms.
Theobald: How necessary is the hydroxyl group ortho to the ketone carbonyl in the 'A' ring? Reduction of the olefinic double bond enhances sweetness in the chalcone glycosides, and introduces more flexibility into the molecule. If the presence of the ortho hydroxyl group is necessary, does it introduce more flexibility into the molecule, or does it introduce more rigidity into the structure by intramolecular hydrogen bonding?
Horowitz: We have not tried the effect of removing an ortho group and replacing it with the hydroxyl, but if you methylate or ethylate the group, it decreases the sweetness and there is a dramatic drop in solubility.
Noble: Have you made the sophorose (2-*O*-β-D-glucopyranosyl glucopyranose) analogue of neohesperidin?
Horowitz: No, we have not. If we live long enough, we might try this!
Shallenberger: The interesting thing about these compounds is the area of the mouth in which they elicit the sweet-taste response.
Horowitz: I should like to emphasise that the sweetness of these hydroxyl groups is not the same as sucrose. The sweetness is felt largely towards the back of the mouth. It is very persistent; 10 minutes later, water tastes sweet, but the onset of sweet response is slow. This may have applications in things like chewing-gum.
Coulson: Having eaten miracle berries at Dr Inglett's suggestion, I found they produced a sweet aftertaste similar to your report for the dihydrochalcones. I wonder whether there is any connection in mechanism?
Horowitz: I do not know: it occurred to me this morning that it would be interesting to taste acid materials after dihydrochalcones, to see if they taste sweet.
Head: Do the dihydrochalcones elicit a response in animals that do not respond to sucrose?
Horowitz: Animals take readily to dihydrochalcone sweetness.
Swindells: The prolonged sweetness of dihydrochalcones could be of value in pharmaceuticals as an aid to masking the protracted, objectionable, *e.g.* bitter, flavour of many oral products.
Horowitz: I agree. This is under consideration.

The Physical Basis of the Sweet Response

L. M. BEIDLER

Department of Biological Science,
The Florida State University, Tallahassee, Florida, USA

ABSTRACT

The sensation of sweetness is quite complex and can best be studied with psychophysical techniques. The physical basis of the sweet response, however, involves the chemical stimulation of a heterogeneous population of taste cells. The characteristics of the individual taste cells and of the nerve fibres that innervate them is the topic of this paper.

Taste cells are notorious for their lack of uniformity of response to a given group of chemical stimuli. The response characteristics vary from one cell to the next. For example, the rank order of sugars according to their stimulating efficiency varies from one nerve fibre to the next. Thus any theory of sweetness based upon molecular structure of the stimulus must be flexible enough to account for such variations. The overall response of a large population of taste cells to sweet stimuli also varies from one species to another. The type of chemical stimulus that elicits sweetness is also species dependent.

A study of the receptor response as a function of concentration indicates that most sweet stimuli bind very weakly to the taste cell surface. The association constant (reciprocal molar concentration) for most sugars is of the order of 10 *and saccharin the order of* 1500 *for man. This binding strength is very much larger for the glycoprotein of miracle fruit which imparts a sweet taste to all normally sour substances. This is presumably due to the binding of the protein to the taste cell surface near the sugar receptor site. It has been proposed that a decrease of pH produces a conformational change in the taste cell membrane which allows the sugars of the glycoprotein to interact with the sugar receptor site.*

Not only are the receptor sites for sugars not identical from one taste cell to the next but there are different receptor sites on the same taste cell.

HOW DO SENSE ORGANS WORK?

The sensory receptor is a cell that is particularly sensitive to one type of stimulus. Its response properties may, however, also depend upon surrounding cells or tissues. For example, the auditory receptors of hair cells are located on the basilar membrane, the movement of which excites the receptors. It is primarily the response characteristics of the membrane that determines the pattern of response of the hair cells to complex sounds. The visual receptors, on the other hand, are more independent of other cells in their environment although the amount of light hitting the visual receptors is controlled by an accessory structure, the pupil. The importance of the environment to the taste cells is not too well known although the enzymes and the buffering capacity of the saliva may play a role in modifying chemical stimuli before they interact with the taste receptors. The pore size may also limit the stimulus interaction but whether this can be controlled is not known.

Sensory receptors are living batteries, being electrically charged. The magnitude of the potential across the taste cell membrane changes with stimulation. It is known that alterations of the physico-chemical properties of the membrane bring about the changes in membrane potential. How the stimulus relates to this alteration is not understood. For example, we know that a light stimulus may bring about conformational changes in the rhodopsin molecules of a visual receptor, but exactly how this is brought about is still obscure. Furthermore, how this leads to a change in membrane potential of the visual rod is unknown. The initial step in stimulation of a taste receptor is thought to be the adsorption of the chemical stimulus molecules to specific receptor sites on the membrane of the taste cell which leads to conformational changes in certain membrane molecules and a resultant depolarisation of the taste cell membrane.[1,2] The approximate mathematical relationship between the strength of stimulus and the magnitude of depolarisation is hyperbolic. This relationship is also either hyperbolic or logarithmic in many other sense organs.

Information concerning the nature of the sensory stimuli must be passed on to nerve fibres which carry the message to higher nervous centres. The intensity of the stimulus is usually coded in the form of frequency of electrical pulses that are transmitted over a sensory nerve fibre. This is linearly related to the magnitude of the sense cell depolarisation. These nerve impulses can be recorded from the nerve using suitable electronic equipment.

In some sense organs, such as the olfactory organ, the receptor and the nerve is part of the same cell. In others, such as the taste bud, the receptor cell is separate and the nerve is stimulated by either a chemical mediator or electrical events that originate in the receptor cell and are transmitted across the cell junction or synapse which it makes with a taste nerve fibre.

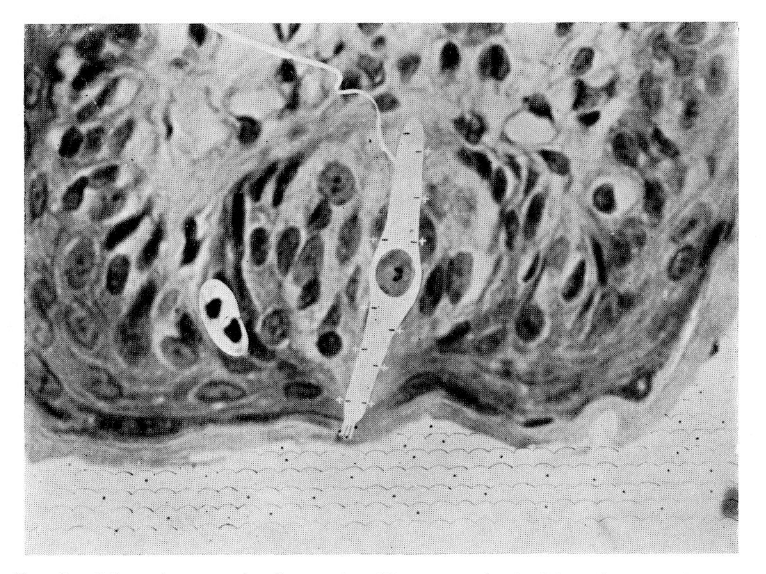

FIG. 1. Microphotograph of a rat fungiform taste bud with a diagramatic taste cell superimposed and a cell undergoing mitotic division to the side.

About 50 taste cells are grouped together to form a taste bud (*see* Fig. 1). Each cell has finger-like projections, microvilli, which are in direct contact with the saliva. Since these cells are constantly insulted by various chemicals placed in the mouth, there must be some method of stabilising the properties of the taste bud. It was discovered that new taste cells could be constantly formed to replace those taste cells that were damaged. The life span of an average taste cell is about 10 days only.[3–5]

A sense receptor does not always signal its message directly to the central nervous system without interaction with other similar receptors. Neighbouring receptors may send nerve processes laterally either to inhibit or facilitate the responses of a given receptor cell. Such interactions have been studied in taste where one nerve fibre may innervate several receptors in a taste bud or even several neighbouring taste buds.[6,7] Thus, there exists considerable processing of the taste information at the level of the taste bud.

TASTE RECEPTOR STIMULATION

Sweet Messages

The taste organs respond to a large variety of molecules. How do they signal the brain that some of these molecules are related to sweetness? The sweet message is not simple. Most single taste cells within the taste bud do not respond to sweet substances alone but also respond to those related to sour, salty, bitter, etc. This was discovered by inserting micro-electrodes into single taste cells and recording the magnitude of electrical depolarisation as various stimuli were applied.[8–11] Similar conclusions were obtained indirectly by recording the electrical messages from single taste nerve fibres.[12,13] The results of such studies also indicated that the cell membranes were quite heterogeneous and contained mosaics of a variety of receptor sites to which molecules could attach.[2] Thus, one site might bind the cations of salt whereas another binds sugars. The relative numbers of these different sites vary from one taste cell to the next.

Are all the sites that bind sweet substances identical? The answer appears to be 'No'. Positive evidence was obtained by Bartoshuk who recorded from single taste nerve fibres of the squirrel monkey in response to several sugars—fructose, sucrose, lactose, dextrose.[14] She found that most responded best to fructose but several favoured sucrose instead. Such differences in response to sugars have been confirmed quantitatively using the response of single rat taste fibres to 1·0 M concentrations of sucrose, xylose and glucose[15] (*see* Table 1). It was also found that the rank order of stimulating efficiency of amino acids varied considerably for rat single taste fibres even though the rank order for the total fibre population remained constant and invariant from individual to individual within the same species (*see* Table 2). If single taste fibres are tested for their responses to a large variety of taste stimuli applied to the tongue, correlations can be calculated for the ability of two different

TABLE 1

The response of four single taste nerve fibres to three sugars

1·0 M	Impulses/10 sec A	B	C	D
Sucrose	199	46	36	30
Xylose	62	27	34	25
Glucose	28	18	38	12

From C. Mistretta, 1970.

TABLE 2

The response of six single taste nerve fibres to three amino acids

0·5 M	Impulses/10 sec A	B	C	D	E	F
DL alanine	112	33	18	12	14	13
glycine	102	44	21	24	51	52
DL threonine	54	16	18	20	35	65

From C. Mistretta, 1970.

TABLE 3

*Correlation coefficient matrix**

Glycine	DL Serine	DL Valine	L Proline	L Tyrosine	
0·33	0·47	0·13	0·18	−0·08	HCl
0·71	0·14	0·46	0·42	0·35	NaCl
0·24	0·45	0·27	0·43	0·05	QHCl
0·51	0·37	0·21	0·21	−0·19	Sucrose
0·65	0·76	0·48	0·37	0·22	DL Alanine
1·0	0·50	0·84	0·68	0·63	Glycine

* The responses of a large number of single taste nerve fibres to ten different taste stimuli, each were recorded and the correlation coefficients of pairs of stimuli are shown.
From C. Mistretta, 1970.

stimuli to excite the same taste fibres. Note in Table 3 the relationship between the amino acids and excitants of other taste qualities. Large species differences in behavioural responses to sugars were also amply illustrated by Kare and Ficken.[16]

Differences in response of single taste nerves might be due to subtle differences in the micro-environment of the various 'sweet' receptor sites on the membranes of the taste cells. If the membrane molecules that make up the taste sites are not over rigid then neighbouring side chains in the vicinity of the site would affect its binding properties.[17]

If there is so much flexibility in the response of nerve fibres, how is the message concerning sweetness transmitted? One must consider the fact that hundreds of nerves are activated at the same time and it is the total activity that is important. This is not too dissimilar to the optic nerves that may be transmitting information about a landscape painting. If one could 'tap in' on but one optic nerve fibre, information might be obtained about the colour and light intensity of but one point in the landscape. If at the same time one could record from a thousand additional fibres, then the elements of the landscape might become apparent. It would take many thousands of fibres in proper perspective before the quality of the landscape painting could be appreciated. Perhaps the sweetness of a pie involves many hundreds of taste fibres.

It should be noted that there are all kinds of sweetness. For example, fructose does not have the identical sweet quality of maltose. Indeed, von Bekesy claims pure sweetness is only obtained with glycine which some of his subjects pronounced as heavenly sweet.[18]

DESIGNING NEW SWEETENERS

In the past flavour chemists synthesised new organic molecules to produce sweeteners of no nutritional value but satisfactory sweetness. Some of these molecules were thought to be non-toxic and were introduced into the market place where they were accepted for several decades. In recent years, the public became diet conscious and the artificial sweeteners were used in more and more foods. Thus, the total *per capita* consumption increased considerably and safe dosage levels were reconsidered. Some sweeteners were prohibited by Federal decree and the search for new sweet molecules was on. In the meantime some of the public became critical of artificial sweeteners. Is there an alternative?

Our laboratory decided to make a different approach to the problem. The fundamental characteristics of the taste receptors were studied with the hope that an understanding of the basic receptor properties would enable us to re-evaluate the conventional approach to sweeteners. Two possibilities were considered. First, the normal properties of the taste cells could be altered, either permanently or temporarily, so that the total flavour of foods could be altered. Second, natural sweeteners, such as sugars, could be joined to other larger and more specific molecules so that the total binding strength of this engineered complex would be materially increased so that the concentration of sugar needed for equivalent sweetness could be materially decreased. Little is known about the molecular construction of the taste cell membrane. Therefore, it would be difficult to engineer a new molecule which would bind in a 'lock and key' manner. It was decided to search the literature for naturally occurring substances that were known to change taste. About a dozen are known and two or three were chosen for study.[19]

The fruit of the Nigerian plant *Synsepalum dulcificum* is known to bring sweetness to all sour foods. In 1963 we purchased several large plants from Smith–Menninger Nursery and 50 small plants from Newcomb Nursery, both in Florida. Two years later there was an adequate and continuous supply for biochemical research. This fruit had been investigated earlier[20] with the hope of identifying the active ingredient, but a solvent could not be found for it. We initially chose saliva and soon learned a bicarbonate buffer was sufficient. Biochemical analysis of a concentrate of the active component indicated it was a glycoprotein. Kurihara joined our efforts in 1966 and successfully purified, isolated and characterised the glycoprotein and indicated it had a molecular weight of about 44,000.[21–23]

Further physiological, biochemical and psychophysical investigation indicated that the protein portion of the molecule specifically attaches to the taste receptor membrane near the normal sweet receptor site. When the pH is made acidic, the receptor molecule changes conformation and allows the sugar portion of the glycoprotein to adsorb to it. In this manner a natural sugar interacts with the sweet receptor site but it is kept in this position by the protein to which it is attached. In other words, the apparent binding strength of the natural sugar has been greatly increased so that sweetness is enhanced without an increase in sugar concentration. The binding of the protein to the human taste cell membrane appears to be quite specific and miracle fruit is found to be inactive when applied to the tongue of the monkey while recording from its taste nerve.[24]

The action of miracle fruit can be tested objectively on humans by recording the electrical activity from the taste nerve as it passes in air through the middle ear. Berries were freeze-dried in our laboratory and tablets containing 40 mg of this material were made for human testing. Likewise lollipops containing a tea made from leaves of *Gymnema Sylvestre* were used to assure uniform coating of the

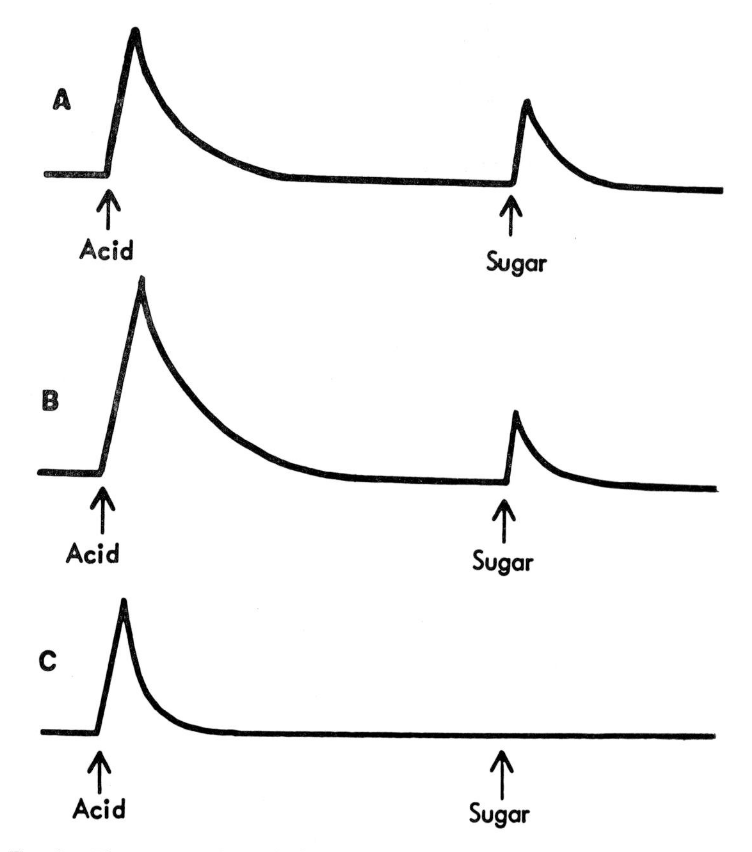

FIG. 2. The response from the human taste nerve to stimulation by 0·02M Citric Acid and 0·5M Sucrose under the conditions: A. control, B. after miracle fruit tablet, C. after miracle fruit tablet followed by gymnemic acid, from Diamant and Zotterman, 1971).

tongue with this sweet-inhibiting material. Both miracle fruit tablets and gymnemic acid lollipops were forwarded to Dr Zotterman in Stockholm for further testing. He and Dr Diamant used these materials to determine their action on sweet and sour stimuli while recording from the taste nerves of patients undergoing middle ear surgery.[24] Results are shown in Fig. 2. Note that the miracle fruit application substantially increases the taste neural activity due to citric acid, but has only a small effect on the response to sucrose. The action of the sweet-inhibiting drug is to eliminate the sucrose response and also that increment of the response to citric acid which was brought forth by the miracle fruit. These experiments agree with the concept that miracle fruit does not interfere with normal acid stimulation but that it stimulates the sweet receptors at low pH values (*see* Table 4).

TABLE 4

	Concentration of citric acid		
	$7 \cdot 5 \times 10^{-4}$ M	$1 \cdot 5 \times 10^{-3}$ M	$2 \cdot 0 \times 10^{-2}$ M
Before taste-modifying protein	Sour	Sour	Very sour
After taste-modifying protein	Sour	Sweet	Very sweet
After taste-modifying protein and gymnemic acid A_1	Sour	Sour	Very sour

MIRACLE FRUIT AS A SWEETENER

From the information we now have, it is apparent that the active ingredient of miracle fruit does not act as a drug, that is, it does not change the taste receptors but merely allows the sugar of the glycoprotein to stimulate the sweet receptor sites of the taste cells in a normal manner. It also acts at low molar concentration (*see* Fig. 3). The strongly attached protein merely restricts the movement of the sugar to the neighbourhood of the receptor site with which it reacts at proper pH. Since the protein attachment to the taste cell membrane is strong, the action of miracle fruit is long lasting (*see* Fig. 4). This characteristic is particularly useful when miracle fruit is used in chewing gum to extend the duration of flavour production.

The sweetness of this glycoprotein is dependent upon its sugars and the duration upon the character of the protein. The magnitude of sweetness of a molecule is dependent upon two factors: the strength of its binding to the receptor site, and the influences on the receptor membrane when it does bind to the site. Once the chemical properties of the above glycoprotein are known, then the molecule

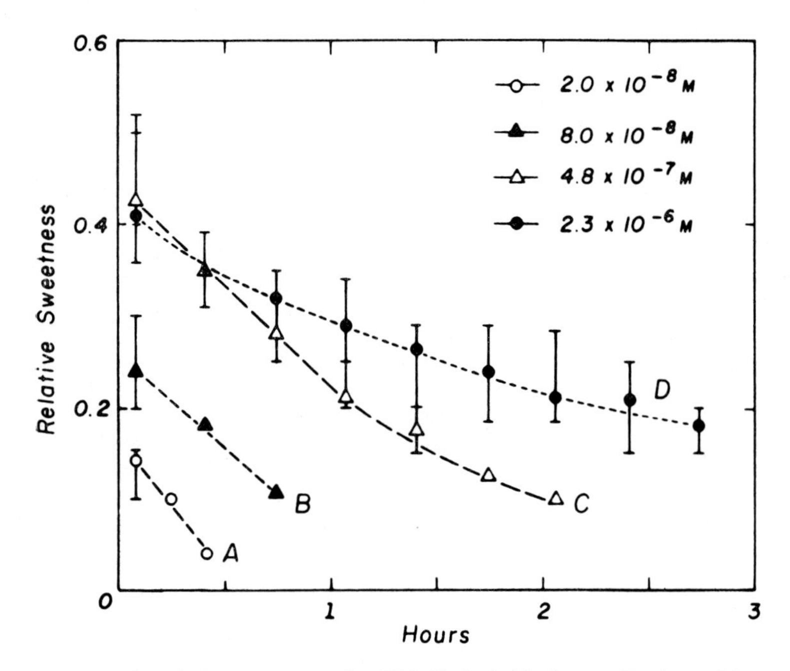

FIG. 3. The relative sweetness of 0·02M Citric Acid after applications of four different concentrations of the isolated miracle fruit glycoprotein at various times after application, (from Kurihara, Kurihara, and Beidler, 1969).

can be re-engineered to make a new molecule with the properties we may desire. For example, the arabinose or xylose sugars might be replaced by a sugar of greater natural sweetness such as fructose. This would decrease the concentration of the new miracle fruit protein necessary for a given sweetness.

The action of this glycoprotein declines with time almost exponentially. Application of this sweetener might be greatly enhanced if this time could be shortened or lengthened according to the intended use. Is this possible? The taste cell membrane contains a net negative charge which may be important in the binding of the protein. Thus, the effective duration of miracle fruit could possibly be altered by blocking some of its reactive groups. It has been shown that the basic residues, histidine and lysine, can be modified by specific reagents and the activity of miracle fruit is lost.[25] On the other hand, tyrosine, tryptophane, carboxyl groups and SH groups are not as important for maintenance of the activity.

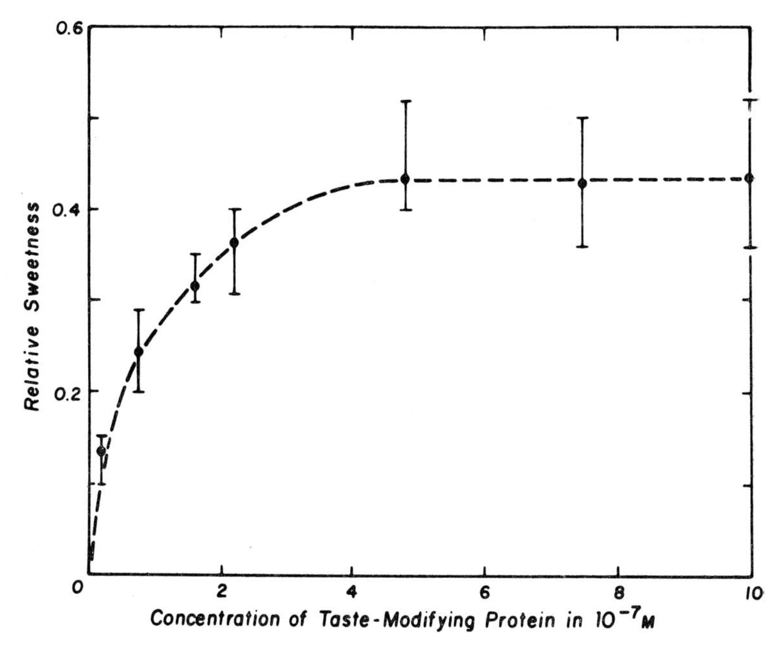

FIG. 4. The relative sweetness of 0·02M Citric Acid after application of various concentrations of isolated miracle fruit glycoprotein (from Kurihara, Kurihara and Beidler, 1969).

CONCLUSION

Knowledge of the basic mechanisms of taste cell stimulation is useful in the engineering of new molecules intended as sweeteners. If oral pollution is to be contained, new and imaginative approaches must be made in the future. A promising approach is the one afforded by characterisation of the active ingredient of miracle fruit. Several other unrelated plants are known to contain similar ingredients and are under investigation. The whole problem of flavour modifiers should be considered more in the realm of biological engineering rather than organic chemistry.

REFERENCES

1. Beidler, L. M. (1954). *J. gen. Physiol.*, **38,** p. 133.
2. Beidler, L. M. (1962). *Progress in Biophysics and Biophysical Chemistry*, Ed. by Butler, J. A. V., Huxley, H. E. and Zirkle, R. E., Pergamon Press, New York, **12,** p. 109.
3. Beidler, L. M., Nejad, M., Smallman, R. and Tateda, H. (1960). *Fedn. Proc.*, **19,** p. 302, Abstract.
4. Beidler, L. M. and Smallman, R. L. (1965). *J. Cell Biol.*, **27,** p. 263.
5. Conger, A. (1969). *Radiat. Res.*, **37**(1), p. 31.
6. Beidler, L. M. (1967). *Olfaction and Taste III*, Ed. by Pfaffmann, C., Rockefeller Univ. Press, p. 352.
7. Miller, I. (1968). Dissertation, Dept. biol. Sci., Florida State University.
8. Kimura, K. and Beidler, L. M. (1961). *J. cell. comp. Physiol.*, **58,** p. 131.
9. Tateda, H. and Beidler, L. M. (1964). *J. gen. Physiol.*, **47,** p. 479.
10. Sato, T. (1969). *Experientia* (*Basel*), **25,** p. 709.
11. Ozeki, M. (1970). *Nature, Lond.*, **228,** p. 868.
12. Pfaffmann, C. (1941). *J. cell. comp. Physiol.*, **7,** p. 243.
13. Fishman, I. Y. (1957). *J. cell. comp. Physiol.*, **49,** p. 319.
14. Pfaffmann, C. (1969). *Reinforcement and Behavior*, p. 215.
15. Mistretta, C. (1970). Dissertation, Dept. of biol. Sci., Florida State University.
16. Kare, M. R. and Ficken, M. S. (1963). *Olfaction and Taste I*, Ed. by Zotterman, Y., Pergamon Press, London, p. 285.
17. Beidler, L. M. and Gross, G. (1971). *Contributions to Sensory Physiology*, Ed. Neff, W. D., Indiana University. In progress.
18. Bekesy, G. von (1964). *J. appl. Physiol.*, **19,** p. 1105.
19. Cohn, G. (1914). *Die Organischen Geschmacksstoffe*, Franz Siemenroth, Berlin.
20. Inglett, G. E. (1965). *J. Agric. Fd. Chem.*, **13**(3), p. 284.
21. Kurihara, K. and Beidler, L. M. (1968). *Science*, **161,** p. 1241.
22. Kurihara, K. and Beidler, L. M. (1969). *Nature, Lond.*, **222,** p. 1176.
23. Kurihara, K., Kurihara, Y. and Beidler, L. M. (1969). *Olfaction and Taste III*, Ed. by Pfaffman, C., Rockefeller University Press, p. 450.
24. Diamant, H. and Zotterman, Y. (1971). *Personal Communication.*
25. Kurihara, K. (1971). *Personal Communication.*

DISCUSSION

Parker: Is there any evidence that some artificial sweeteners stimulate the nerve endings of the sweetness receptor cells directly by transport across the membrane? For example, the simple aromatic synthetic sweeteners, chloroform or ethanol.
Beidler: We have not tried many; none of the sweet stimuli tried passed through the membrane. There is no evidence for transport across the membrane; if it did occur, the response would not be sweet but, possibly, metallic. The response to salt is about 12 msec.
Allen: Is ethanol unique in its action of passing through the membrane?
Beidler: No. Other alcohols exhibit the same effect, but to different degrees.
Allen: Have you tried essential oils?
Beidler: No.
Coulson: We heard yesterday Professor Shallenberger in discussion on Professor Horowitz's paper say that dihydrochalcones affected cells in the roof of the mouth. Just over a year ago Professor Morley Kare reported in London the very rapid transfer of glucose through the roof of the mouth into the brain. Is there any connection?
Beidler: It is doubtful, as different taste buds are involved.
Cristofaro: Does any quantified relationship exist between age of the taster and perception?
Beidler: There is a lot of folklore associated with this question, but I do not know of any scientific evidence.
Cristofaro: Concerning the preference of babies for sweet food, do they have the same number of receptors as adults, and are these receptors fully developed?
Beidler: The receptors are fully developed by five foetal months. At birth, the baby probably has more taste buds than you do, but starts to lose them almost immediately. If, after birth, an infant is offered a choice between bitter and sweet, by painting one of the mother's nipples with a bitter substance, it will choose the bitter. It will do this for a short time and then change to the sweet milk.
Coulson: Supplementary to the question on baby response to sweetness, the Institute of Mother and Child, Warsaw, some three years ago, carried out preference tests on babies. They showed preference for spicy and piquant flavours, not sweet. This perhaps relates more to Dr Watson's paper yesterday.
Gmunder: Some sugars have a sweet taste followed by an irritant 'scratchy' sensation in the throat. Does Professor Beidler have an explanation of this phenomenon? Is it true that one experiences a sweet response at the front of the tongue and a different aftertaste at the back?
Beidler: There is a different magnitude of response at the front and the back. The sweet response is greater at the front of the tongue whereas the response to bitter is the opposite. This is due to differences in transport rates.
Spence: Is there any evidence to suggest that the individual taste cells change their sensitivity characteristics during their life time?

Beidler: I wish I knew about that. This might be their mechanism of coding.
Clark: You say that taste cells live for 10 days, each having different responses to acidity, sweetness etc. Does this mean that taste perception can change from day to day?
Beidler: The total population of taste cells is constant and the total nerve response gives a constant response to the same stimuli.
Kropp: It is experienced from practical results that a mixture of different sugars is sweeter than could be expected from the sweetness of the single components. Can you give a physiological explanation for this phenomenon?
Beidler: I do not know. If, for example, there are two amino acids mixed together, it is difficult without a model, or a theory, to find out if the two stimuli operate simultaneously.
Crampton: How can one taste a physiological saline solution, when, surely, the taste buds themselves are presumably exposed to identical solutions (saliva and extra-cellular fluid) containing salt, amino acids, acetate, etc.?
Beidler: The membrane prevents migration of the salts to the nerve which is situated on the other side of the membrane, away from the microvilli.
Grenby: Is there any explanation of the bitter after-taste which some people notice from saccharin?
Beidler: I do not know. Perhaps because with things that are predominantly sweet, you are stimulating both bitter and sweet.
Macdonald: Is it true that only water soluble substances can be tasted?
Beidler: Yes.
Adams: Miracle berry reacts with acidic groups; if one takes fructose or saccharin after dosage with miraculin, is the effect additive or compounded?
Beidler: No, you taste the appropriate sweetener.
Pangborn: Do identical twins, animal or human, have similar taste sensitivities?
Beidler: Work has been done, but I cannot remember the answer.
Mears: If only water soluble materials can be tasted, how do you account for the effect of essential oils? At high concentration they can be astringent.
Beidler: Essential oils will be tasted on the free nerve endings and not on the taste buds. For the latter to be stimulated, the material must be water and lipid soluble.

The Chemical Basis of Sweetness in Model Sugars

G. G. BIRCH and C. K. LEE

National College of Food Technology,
University of Reading, Weybridge, Surrey, England

ABSTRACT

The chemical basis of sweetness is investigated on the assumption that it is a stereochemical effect involving the binding of particular types of molecule to the taste bud protein. Only by examining a large number of such molecules will chemical patterns which govern the strength of the binding begin to emerge. The most suitable model sugars for analysing the phenomenon of sweetness appear to be glycosides and non-reducing sugars. In particular the disaccharide, α, α-trehalose (mushroom sugar) is ideal as a starting point, and many of its derivatives have consequently been tested organoleptically. Also of value as stable well-defined molecules are the cyclitols which have varying degrees of sweetness depending on the molecular geometry of their rings and glycol groupings. Explanations are advanced for some of these differences and for those between non-reducing sugar derivatives. The well-known decrease in sweetness with increasing molecular weight could be explained by assuming that only one sugar residue in each oligosaccharide is actually involved in binding to the taste bud protein. One experiment is described which supports this hypothesis. It is, furthermore, quite feasible that only part of a saporific sugar residue is specifically responsible for eliciting the sweet response.

INTRODUCTION

It is not known whether diffusion of saporific substances across taste bud cell membranes is a rate-determining step governing the intensity of the sweet response, or whether direct binding of saporific molecule with taste bud protein is the criterion of magnitude. The latter possibility has been chosen as a working hypothesis in the case of sugars and related substances, largely because of evidence already published by Shallenberger *et al.*,[1-10] to account for organoleptic differences caused by configurational changes (Fig. 1), and because the field of carbohydrate-protein binding is already being explored in many other physiological aspects.

Marked organoleptic effects caused by single configurational changes as demonstrated in Fig. 1 can simply be explained in terms of molecular geometry, but if such a system of binding as postulated

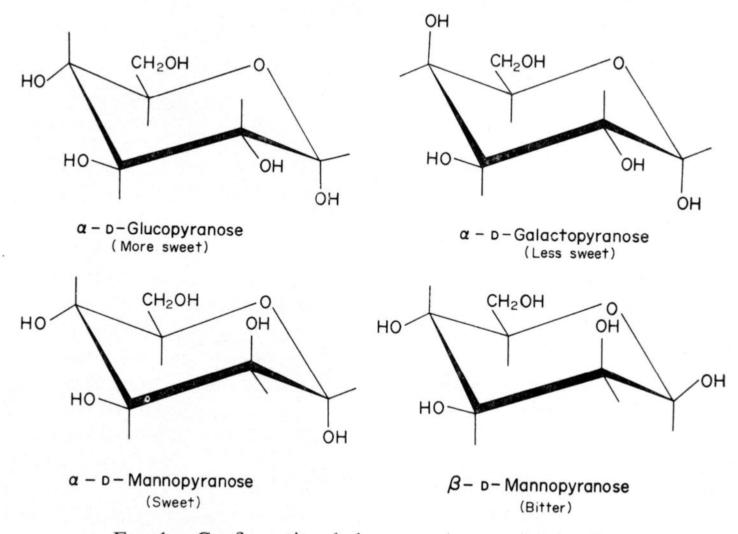

FIG. 1. Configurational change and organoleptic effect.

by Shallenberger is operating it is still not known whether one or several receptor molecules are involved. Furthermore the mechanism by which the protein, once bound, transmits the signal to the nerve fibre is still unexplained, although several theories in the general field of sensation have been advanced. Randebrock,[11] for example, believes that different odoriferous molecules may bind to a common perceptor protein in the form of an α-helix stabilised by hydrogen bonds. The mode of binding (dependent on molecular geometry) then governs the type of oscillation which is transmitted down the chain of hydrogen bonds stabilising the α-helix, and is then transmitted to the nerve.

Since each saporific sugar has a number of different α-glycol groupings, each of which meets Shallenberger's requirements of a geometrically suitable AH,B system there is clearly a need to test the

Shallenberger theory by modifying these α-glycol groupings in various ways. For example, selected oxygen atoms can be eliminated to form deoxy sugars, and water molecules can be eliminated by condensation and bridging reactions to form anhydro derivatives. By these means it should be possible not only to test the Shallenberger theory but also to decide which particular α-glycol groups constitute the specific sweet-eliciting function in sugar molecules, presumably for geometrical reasons. Shallenberger's theory clearly has many attractions as a simple explanation of sweetness. However, the evidence he has collected has mostly resulted from tests with reducing sugars, which undergo rapid isomerisation in solution, and are not therefore good substrates. Any research programme based on the relationship between stereochemical structure and organoleptic effect must therefore depend entirely on the correct choice of a model sugar.

GLYCOSIDES AS MODEL SUGARS

One way of preventing the difficulties which are experienced with reducing sugars is to substitute the free reducing centre with an alkyl or aryl group. The products are then glycosides and have the advantage that they do not isomerise when dissolved in neutral aqueous fluids. Since they also frequently adopt energetically preferred and well-defined conformations they would appear to be useful models

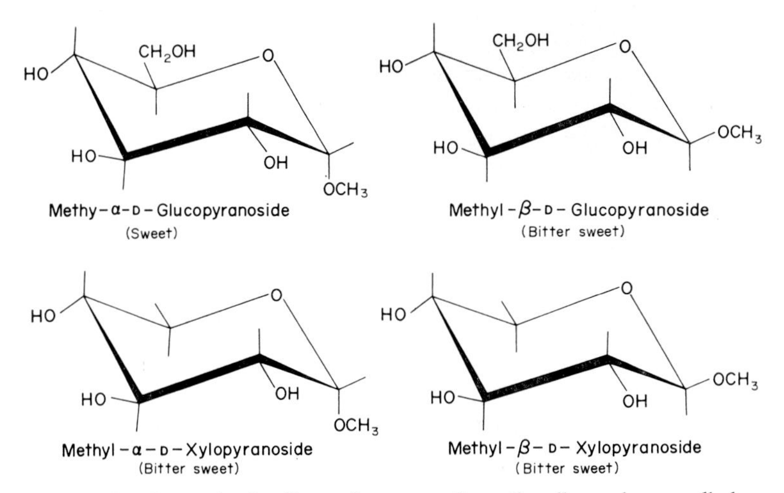

FIG. 2. Organoleptic effects of some conformationally analogous alkyl glycosides.

for investigating the chemical basis of sweetness. Many glycosides and their derivatives can in fact yield valuable information as can be seen in Tables 1–6 below. Unfortunately this approach is complicated by the discovery that whereas aryl glycosides are often too intensely bitter for testing, alkyl glycosides are in many cases bitter-sweet (Fig. 2).

Of the compounds illustrated in Fig. 2 only methyl α-D-glucopyranoside appears to be free of the complicating bitter taste. Even this glycoside, however, appears to be slightly bitter to some taste panellists. The aglycone is therefore frequently an unnatural complicating moiety, and model substances which do not contain aglycones might be expected to give a clearer understanding of the chemical basis of sweetness.

OLIGOSACCHARIDES AS MODEL SUGARS

Oligosaccharides offer certain obvious advantages as models for organoleptic studies. Thus they are often found to be naturally occurring substances devoid of aglycone substituents, sweet and free of complicating taints. Unfortunately, however, most oligosaccharides are reducing sugars, and in any case when they reach the molecular weight of trisaccharide and above they are usually tasteless (or practically so). There are very few non-reducing sweet sugars available but of those which are, sucrose presents the first and most obvious test case (Fig. 3).

Sucrose is sweeter than most sugars and its chemical modification has already been widely studied.[12,13] Such studies could indeed lead to a better understanding of the chemical basis of sweetness but are made difficult by the fact that the two halves of the sucrose molecule are dissimilar. Hence chemical modification may proceed differently in each sugar residue and the products are difficult to analyse spectrophotometrically (*e.g.* by NMR), because of overlapping of peaks, signals, etc.

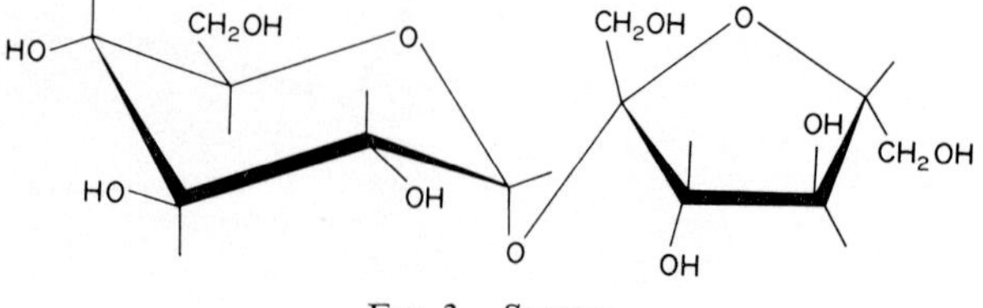

FIG. 3. Sucrose.

What really is needed is a symmetrical, sweet, non-reducing disaccharide. Evidently there is only one such sugar known which is both readily available commercially and well understood conformationally, namely α,α-trehalose, or mushroom sugar [14–24] (Fig. 4).

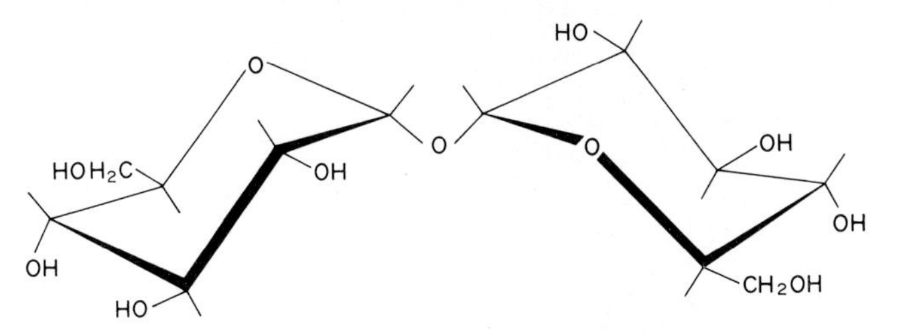

FIG. 4. α, α-Trehalose (Mushroom sugar).

This conclusion is an extremely fortunate one for our laboratory at the National College of Food Technology (Reading University) where the chemistry of the sugar has been extensively investigated for some years now in collaboration with Queen Elizabeth College (London University). With regard to its chemical modification we therefore already have a fund of experience on which to draw. From the results of derivatives already tested (Tables 1–6) (which include derivatives of the analogue methyl α-D-glucopyranoside) it is now possible to draw some tentative conclusions about the chemical reasons for the phenomenon of sweetness.[25,26]

TABLE 1

Organoleptic properties of free reducing sugars (monosaccharides)

Sugar	Sweetness	Bitterness
1. D-Ribose	S	0
2. D-Arabinose	S	0
3. L-Arabinose	S	0
4. D-Xylose	S	0
5. D-Glucose (α)	S	0
6. D-Galactose	S	0
7. D-Mannose (α)	S	0
8. D-Mannose (β)	S	B
9. L-Rhamnose	S	0
10. L-Fucose	S	0
11. D-Quinovose	S	0 or tr. ?

TABLE 2

Organoleptic properties of free reducing sugars (di-, tri- and tetra-saccharides)

Sugar	Sweetness	Bitterness
1. Maltose	S	0
2. Cellobiose	0	0
3. Nigerose	S	0
4. Laminaribiose	?	?
5. Kojibiose	?	?
6. Sophorose	?	?
7. Isomaltose	S	0
8. Gentiobiose	S	B
9. Lactose	S	0
10. Maltotriose	tr.	0
11. Panose	tr.	0
12. Raffinose	0	0
13. Maltotetraose	0	0
14. Stachyose	?	?

TABLE 3

Organoleptic properties of simple glycosides

Glycoside	Sweetness	Bitterness
1. Methyl-α-D-glucopyranoside	S	0
2. Methyl-β-D-glucopyranoside	S	B
3. Methyl-α-D-xylopyranoside	S	B
4. Methyl-β-D-xylopyranoside	S	B
5. Methyl-α-D-galactopyranoside	S	0
6. Methyl-β-D-galactopyranoside	S	0
7. Methyl-α-D-mannopyranoside	S	B
8. Methyl-α-D-quinovoside	tr.	0
9. Methyl-α-L-rhamnoside	tr.	tr.

TABLE 4

Organoleptic properties of simple glycoside derivatives

Glycoside derivatives	Sweetness	Bitterness
1. Methyl-4,6-benzylidene α-D-glucopyranoside	0	B
2. Methyl-4,6-benzylidene α-D-altropyranoside	0	BB
3. Methyl-3-deoxy-β-L-erythro-pentopyranoside	0	B
4. Methyl-α-L-3-acetamido-3-deoxy-quinovoside	tr.	tr.
5. Methyl-3-acetamido-2,3-dideoxy-α-D-glucoside	tr.	0
6. Methyl-3-acetamido-2,3-dideoxy-4,6-benzylidene-α-D-glucoside	0	0
7. Methyl-3-acetamido 2,3-dideoxy-α-D-quinovoside	tr.	tr.
8. Methyl-3-acetamido-2,3-dideoxy-α-D-fucoside	?	?
9. Methyl-3-acetamido-3-deoxy-α-L-fucoside	tr.	tr.
10. Methyl-2-acetamido-2,4,6-trideoxy-α-D-glucoside	0	tr.
11. Methyl-3-amino-3,6-dideoxy-α-L-glucoside	tr.	tr.
12. 1,6-Anhydro-3-dimethylamino-3-deoxy-β-D-glucoside	tr.	0
13. 1,6-Anhydro-3-deoxy-3-dimethylamino-β-D-altroside	tr.	0
14. 1,6-Anhydro-3-acetamido-3-deoxy-β-D-glucoside	tr.	0
15. 1,6-Anhydro-β-D-glucoside (levoglucosan)	tr.	(astringent) B
16. Sedoheptulosan	S	0

TABLE 5

Organoleptic properties of trehalose derivatives and non-reducing sugar analogues

	Sweetness	Bitterness
1. α,α-Trehalose	S	0
2. α,α-'*Galacto*' trehalose	S	0
3. 6,6′-Dideoxy-α,α-trehalose	tr.	0
4. *O*-α-D-xylopyranosyl-α-D-xylopyranoside	S	0
5. Sucrose	SS	0
6. 2-Deoxyribofuranosyl-2-deoxyribofuranoside	S	tr.
7. 4,6:4′,6′-Di-*O*-ethylidene-α,α-trehalose	0	tr.
8. 4,6:4′,6′-Di-*O*-benzylidene-α,α-trehalose	0	0
9. 4,6:4′,6′-Di-*O*-benzylidene 2-*O*-tosyl-α,α-trehalose	0	0
10. 4,6:4′,6′-Di-*O*-benzylidene-α,α-'altro'-trehalose-2,3:2′,3′-di-epoxide	0	0
11. 4,6:4′,6′-Di-*O*-benzylidene-2-amino-'*altro*'-trehalose	tr.	0
12. 4,6:4′,6′-α,α-'*Allo*' trehalose 2,3:2′,3′-di-epoxide	0	0
13. 6-*O*-Mesyl-α,α-trehalose	S	tr.
14. 6,6′-Di-*O*-mesyl-α,α-trehalose	tr.	B
15. 6-*O*-Tosyl-α,α-trehalose	0	B
16. 6,6′-Di-*O*-trityl-α,α-trehalose	0	0
17. 3,6-Anhydro-α,α-trehalose	S	0
18. 3,6:3′,6′-Dianhydro-α,α-trehalose	0	0
19. 2,3:2′,3′-Tetra-*O*-benzoyl-α,α-trehalose	0	0
20. Octa-*O*-benzoyl-α,α-trehalose	0	0
21. 6,6′-Diazido-6′,6′-dideoxy-4,4′-dimesyl-tetra-*O*-benzoyl-α,α-trehalose	0	0
22. 2,2′-Diazido-2,2′-dideoxy-α,α-'*altro*'-trehalose	0	0
23. 4,4′-Diacetamido-4,4′-dideoxy-α,α-'*galacto*'-trehalose	0	0
24. 4,6:4′,6′-Tetraacetamido-α,α-'*galacto*'-trehalose	tr.	0

TABLE 6

Organoleptic properties of acetates

Acetate	Sweetness	Bitterness
1. Tetra-*O*-acetyl-α-D-xylose	0	BB
2. Penta-*O*-acetyl-α-D-glucose	0	BB
3. Penta-*O*-acetyl-β-D-glucose	0	BB
4. Methyl-tetra-*O*-acetyl-α-D-glucoside	0	BB
5. Penta-*O*-acetyl-α-D-galactose	0	BB
6. Methyl-tetra-*O*-acetyl-α-D-galactoside	0	BB
7. Methyl-α-L-3-acetamido-3-deoxy-2,3-di-*O*-acetyl-fucoside	0	B
8. Octa-*O*-acetyl-β-D-maltose	0	BB
9. Octa-*O*-acetyl-β-D-cellobiose	0	B
10. Octa-*O*-acetyl-α,α-trehalose	0	BB
11. Octa-*O*-acetyl-sucrose	0	BB
12. Octa-*O*-acetyl-β-D-lactose	0	BB
13. Octa-*O*-acetyl-α,α-'*galacto*' trehalose	0	BB

(i) Shallenberger's sweetness hypothesis still appears to be a feasible explanation for the organoleptic differences which are caused by chemical modification.

(ii) The fourth hydroxyl group of glucopyranoside structures appears to be of unique importance in eliciting the sweet response—possibly donating the proton as the AH of Shallenberger's AH,B system.

(iii) Perhaps surprisingly the primary alcohol group does not appear to be of great importance.

(iv) Substituton of acetyl or azide groups in sugar molecules confers intense bitterness. Substitution of benzoyl groups on the other hand causes tastelessness, probably due to lack of solubility.

Conclusion (ii) was also reached by Evans[27] in taste-preference studies with blowflies.

QUANTITATIVE ASSESSMENT OF SWEETNESS

Any quantitative assessment of sweetness must of course depend on statistically reliable use of taste panels. Such work is laborious and is of necessity restricted to solutions which are not complicated by significant levels of non-sweet tastes, synergistic or depressant effects. It is nevertheless possible in certain instances to select substances related in their molecular architecture and to compare their individual sweetness intensities. This has been done for mushroom sugar (α,α-trehalose)[28] and its analogue methyl α-D-glucopyranoside, and the result is that on a molar basis their sweetnesses are the same (Fig. 5).

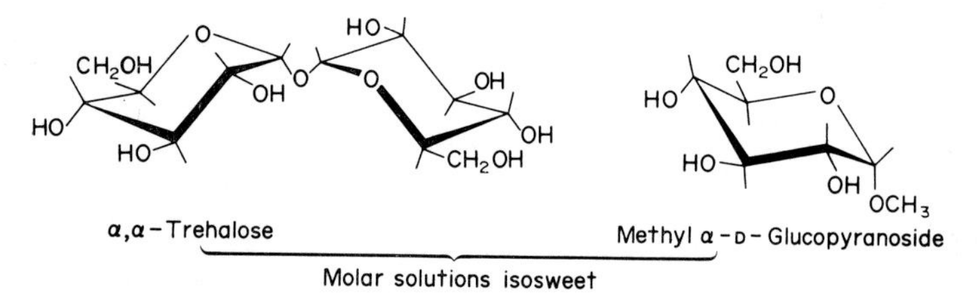

FIG. 5. Sweetness intensity in model sugar analogues.

Hence it appears that if Shallenberger's proposed mechanism is operating, only one-half of the trehalose molecule is involved in binding to the taste bud protein, the other half being excluded presumably for steric reasons. If this result also applies to the disaccharide sucrose, it is clearly the fructofuranose residue which is responsible for the sweetness in this molecule.

1-DEOXY SUGARS

A new method of avoiding the difficulties encountered with reducing sugars, glycosides or oligosaccharides as substrates for organoleptic work is to eliminate the oxygen atom at their free reducing centres. Such substances can be prepared by dehydration of open chain hexitols and are sweet in accordance with our earlier prediction that the fourth hydroxyl of glucopyranosides is the one of greatest significance. This supports the idea that the first like the sixth hydroxyl group is not of great importance in eliciting sweet response (Fig. 6).

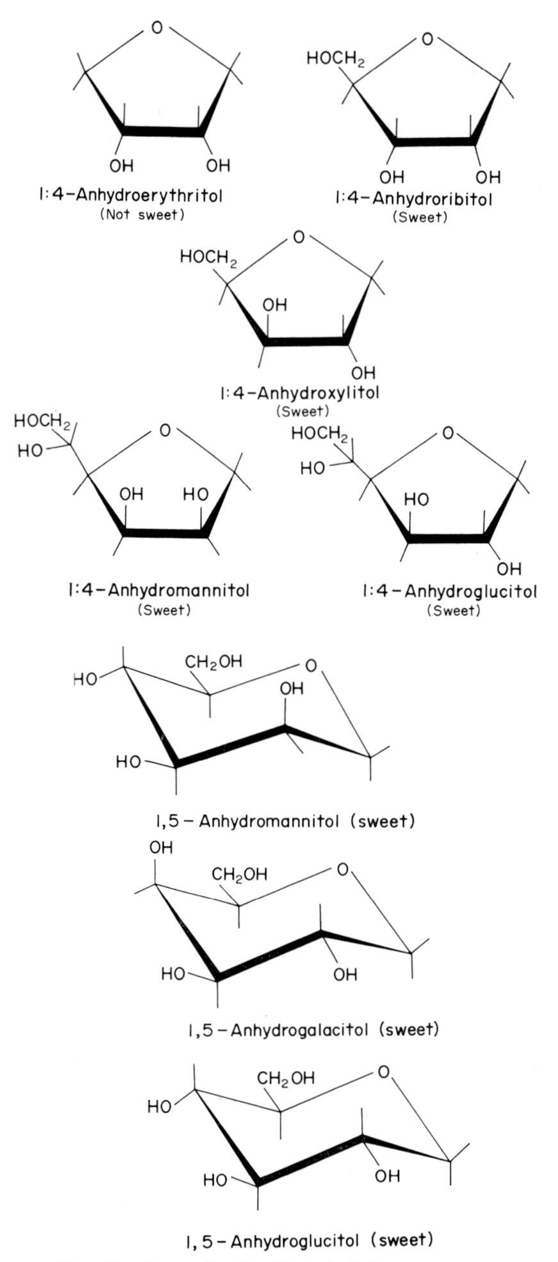

FIG. 6. Organoleptic effects in l-deoxy sugars.

CHANGES IN RING SIZE AND SHAPE

It is quite possible that change in ring size or conformation may, by affecting molecular geometry, also cause changes in organoleptic effect. Since sugar molecules tend to adopt preferred conformations it is not possible to alter conformation without substitution or bridging reactions. One such change has been achieved[29] with the formation of the symmetrical 3,6-dianhydro derivative of α,α-trehalose (Fig 7).

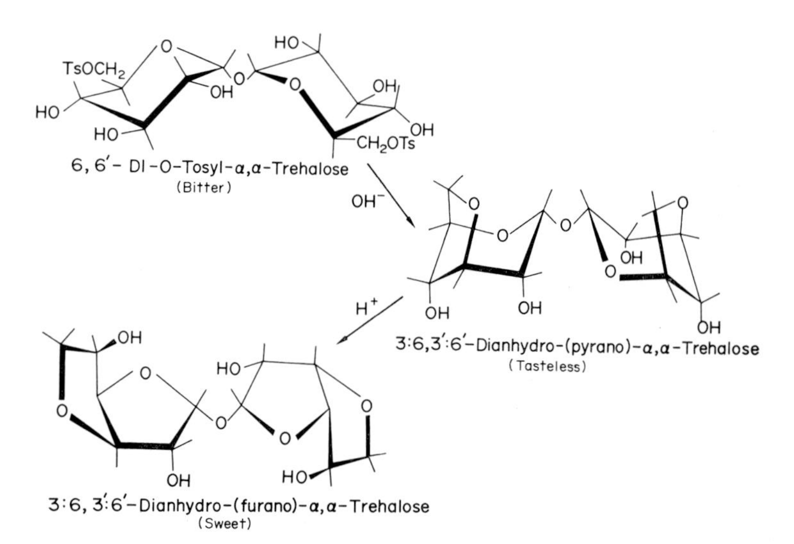

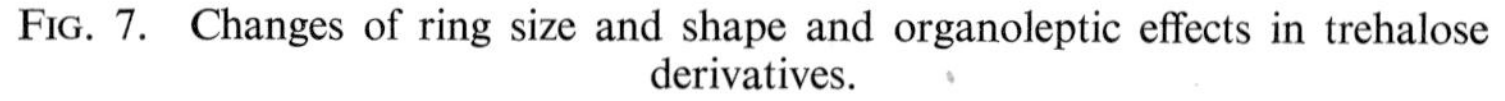
FIG. 7. Changes of ring size and shape and organoleptic effects in trehalose derivatives.

In the pyrano-derivative, change in conformation from the more stable to the less stable chair-form has occurred, whereas the final stage in the reaction results in ring contraction to the furanose form. The sweetness of the dianhydro-furano trehalose poses an interesting problem. The molecule contains no α-glycol groupings, nor does it contain any AH,B system of the usual interatomic (A–B) distance of 2·8–2·9 A. Furthermore it conflicts with references in Shallenberger's papers to 'non-sweet furanose forms'. However, the molecule is an analogue of 1,4:3,6-dianhydro sorbitol, a substance which is readily available commercially under the trade name 'Isosorbide', and which is also sweet (Fig. 8).

These substances are 1-deoxy sugars.

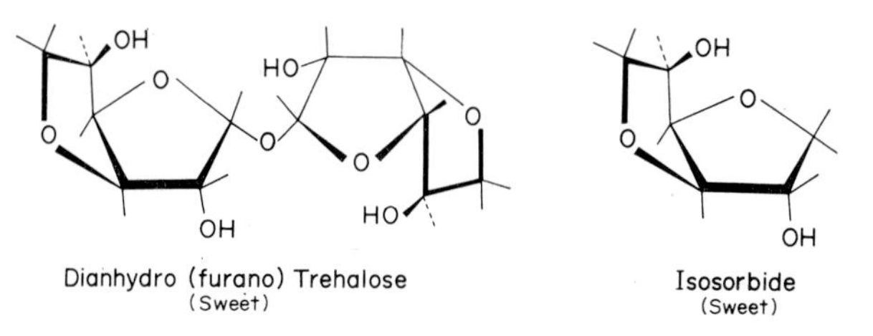

FIG. 8. Sweet anhydro furanose analogues.

SIMPLER MODEL STRUCTURES

An exhaustive analysis of all different possible geometrical arrangements of α-glycol groupings would alone be useful in assessing the chemical reasons for organoleptic phenomena. For this reason it is possible to proceed with organoleptic studies of molecules which are simpler than sugars. Replacement of the ring oxygen atom of sugars with a methylene group gives rise to the cyclitols and a number of these model substances are already available to us, largely due to the notable work of Professor S. J. Angyal (New South Wales) and Professor G. E. McCasland (San Francisco). Cyclitols are often easy to define conformationally and remain stable in neutral aqueous solution. Many of them have the added advantage of occurring naturally—*e.g.* inositols and quercitols (oak sugars)—and offer the exciting possibility of finding a new natural sweetening agent as a food additive. One early finding is that 1,2:4,5 cyclohexane tetrol is not sweet even though it contains two *gauche* α-glycol groupings, each of which therefore meets Shallenberger's criterion for sweetness (Fig. 9).

It appears very probable that lack of sweetness in the above

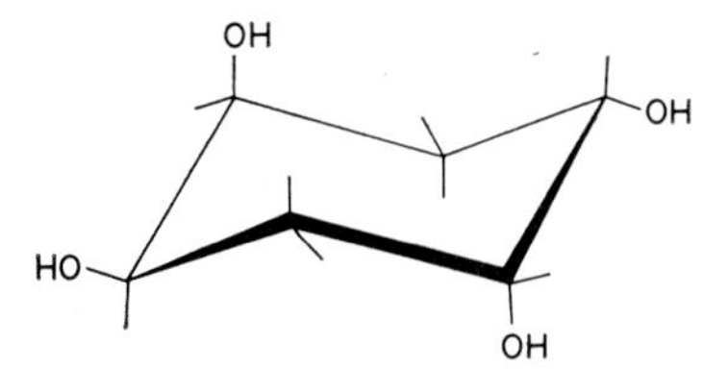

FIG. 9. 1,2:4,5 Cyclohexane tetrol (tasteless).

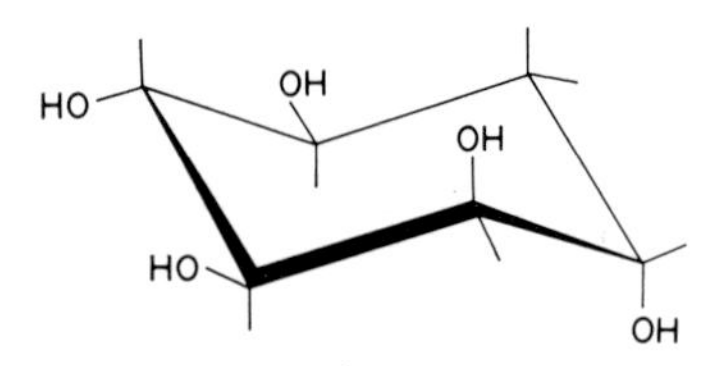

FIG. 10. D-Protoquercitol (an oak sugar) (tasteless).

molecule is due to steric interference by one or both of the axial hydroxyl groups, thus preventing binding to the taste bud protein. Similarly D-protoquercitol (one of the naturally occurring oak sugars) is not sweet even though it contains five hydroxyls and three α-glycol groupings meeting Shallenberger's requirements for sweetness. In this case the trans-diaxial grouping probably causes steric hinderance to binding (Fig. 10).

The racemic mixture D,L-viboquercitol, on the other hand, is quite sweet and if only one enantiomorph is responsible for the effect it is probably the D-isomer because it is an analogue of α-D-glucopyranose having a methylene group replacing the ring oxygen atom and an equatorial hydroxyl replacing the primary alcohol group at C5 (Fig. 11).

An attempt is now being made to obtain the separate enantiomorphs of viboquercitol. Studies with such molecules and a range of other conformationally defined cyclitols which are under way may

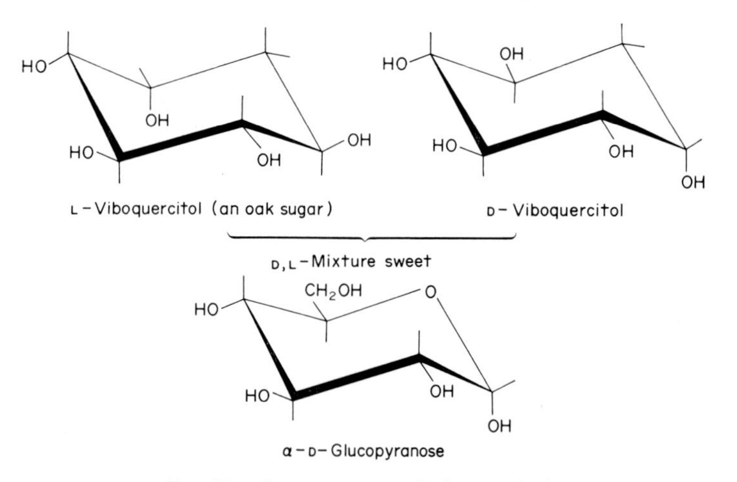

FIG. 11. Sweet response of viboquercitol.

reveal the exact spatial requirements for binding to occur with taste bud proteins, hence the sequence of amino acids at a specific binding site might be predicted. However, it is possible that the mode of binding may alter with different substrates due to induced conformational distortion, as is already known to occur for example with certain cholinergic drugs.[30]

CONCLUSION

Qualitative and quantitative organoleptic assessment of simple, well-defined, conformationally stable substances appears to be a logical starting point for an investigation into the chemical basis of sweetness. A programme of synthesis of model sugars, sugar analogues and derivatives has therefore begun, and of particular interest as model structures are certain sweet naturally occurring substances such as mushroom sugar and oak sugar. Modern techniques of organic chemistry and spectral analysis have facilitated the lines of approach to the problem, and some clear indications of specific geometrical requirements for sweetness have already been arrived at.

Moncrieff[31] has recently stated that, after trying hard for thirty years to find a relationship between one fundamental taste and chemical constitution, in his view 'there is none'. This pessimistic conclusion from an eminent worker in the field of chemical senses is not borne out by preliminary work begun in this laboratory.

ACKNOWLEDGEMENT

We thank J. Sainsbury Ltd, for a grant in aid of sweetness research. Professor G. E. McCasland (San Francisco), Professor S. J. Angyal (New South Wales), and Professor R. Barker (Iowa) are thanked for their interest and kind gifts of samples.

REFERENCES

1. Shallenberger, R. S. (1963). *J. Fd. Sci.*, **28,** p. 584.
2. Shallenberger, R. S. (1964). *Agric. Sci. Rev.*, **2,** p. 11.
3. Shallenberger, R. S. (1964). *New Scient.*, **407**(3), p. 569.

4. Shallenberger, R. S., Acree, T. E. and Guild, W. E. (1965). *J. Fd. Sci.*, **30,** p. 560.
5. Shallenberger, R. S. (1966). *Frontiers in Food Res. (Proc. Symp., Cornell Univ., Ithaca, N.Y.)*, p. 45.
6. Shallenberger, R. S. and Acree, T. E. (1967). *Nature, Lond.*, **216**(4), p. 480.
7. Shallenberger, R. S. (1968). *Frontiers in Food Res. (Proc. Symp., Cornell Univ., Ithaca, N.Y.)*, p. 40.
8. Shallenberger, R. S., Acree, T. E. and Lee, C. Y. (1969). *Nature, Lond.*, **221,** p. 555.
9. Shallenberger, R. S. and Acree, T. E. (1969). *J. agric. Fd. Chem.*, **17**(4), p. 701.
10. Henkin, R. I. and Shallenberger, R. S. (1970). *Nature, Lond.*, **227,** p. 965.
11. Randebrock, R. E. (1968). *Nature, Lond.*, **219,** p. 503.
12. Levi, I. and Purves, C. B. (1949). *Adv. Carbohyd. Chem.*, **4,** p. 1.
13. Wiggins, L. F. (1949). *Adv. Carbohyd. Chem.*, **4,** p. 293.
14. Birch, G. G. (1963). *Adv. Carbohyd. Chem.*, **18,** p. 201.
15. Birch, G. G. (1965). *J. Chem. Soc.*, p. 3489.
16. Birch, G. G. (1966). *J. Chem. Soc.* (C), p. 1072.
17. Birch, G. G. and Cowell, N. D. (1967). *Carbohyd. Res.*, **5,** p. 232.
18. Birch, G. G. and Richardson, A. C. (1968). *Carbohyd. Res.*, **8,** p. 411.
19. Birch, G. G. and Richardson, A. C. (1970). *J. Chem. Soc.* (C), p. 1035.
20. Ali, Y., Hough, L. and Richardson, A. C. (1970). *Carbohyd. Res.*, **14,** p. 181.
21. Birch, G. G., Lee, C. K. and Richardson, A. C. (1970). *Carbohyd. Res.*
22. Hough, L., Munroe, P. and Richardson, A. C. (1971). *J. Chem. Soc.* In press.
23. Hough, L., Richardson, A. C. and Tarelli, E. (1971). *J. Chem. Soc.* In press.
24. Hough, L., Richardson, A. C. and Tarelli, E. (1971). *J. Chem. Soc.* In press.
25. Birch, G. G. (1970). Paper presented on Trehalose Day, in connection to the V International Carbohydrate Symposium, Paris.
26. Birch, G. G., Lee, C. K. and Rolfe, E. J. (1970). *J. Sci. Fd. Agric.*, **21,** p. 650.
27. Evans, D. R. (1963). *Olfaction and Taste*. Ed. by Zotterman, Y. Int. Symp. Series, Vol. 1, p. 165. Oxford, Pergamon Press.
28. Birch, G. G., Cowell, N. D. and Eyton, D. (1970). *J. Fd. Technol.*, **5,** p. 277.
29. Birch, G. G., Lee, C. K. and Richardson, A. C. (1971). *Carbohyd. Res.*, **16,** p. 235.
30. Portoghese, P. S. (1970). *A. Rev. Pharmac.*, **10,** p. 51.
31. Moncrieff, R. W. (1970). *Flav. Ind.*, p. 583.

DISCUSSION

Harborne: 1. It is quite clear from your paper and also the paper of Dr Horowitz that the structural requirements for bitterness are very closely related to those for sweetness. Have theories similar to the Shallenberger theory been put forward to explain bitterness?

2. The substitution of the ring-oxygen of model sugars by sulphur might be expected to have a dramatic effect on taste properties. Have the organoleptic properties of thio-sugars been tested?

Birch: 1. The only one is the Kubota theory which is based on Shallenberger's requirements of an AH,B system, but there is not much evidence in support of this.

2. I have not tasted any thio-sugars or nitrogen-sugars.

Shallenberger: Thio-D-glucose is slightly sweeter than glucose, but also slightly bitter.

Noble: Your own work and that of Professor Shallenberger is concerned with looking at sugars, the interatomic distances between hydroxyl-groups, whether they are gauche or skew and whether any hydrogen bonding exists. I expect you are familiar with Anderson's work in Scandinavia in which he examined the response of the *chorda tympani* to various sweet substances placed on the tongue. He found that the magnitude of the response is related to the solubility of the particular sugar. I wonder whether you, or Professor Shallenberger, have determined the solubilities of your sugars and tried to relate them to the magnitudes of their sweetnesses?

Birch: I have not. And the work with solid crystals has been done on a very rough basis. I do not think the distinctions between sweetnesses will enter into the observations under these circumstances.

Young: Has Dr Birch investigated the differences in taste between the two fructosyl glucoses in which the 3- and 4-hydroxyl groups are glycosided respectively? It will be interesting to try also the various glucosyl fructoses (turanose and maltulose for example).

Birch: I have not. I tend to avoid these sugars because they are reducing sugars.

Stacey: Has the speaker built atomic models of the cyclitols and compared the sweet ones with the sweet sugars? There are significant shape differences between the various cyclitols and the various sugars, and it seems remarkable that the same receptor molecule could accept such a wide variety of molecules to give the sweetness phenomenon.

Birch: We do use models and it is only by using a great number of these models that we shall build up a picture of the geometrical arrangements necessary to elicit the reasons for sweetness.

Spencer: In sweetness intensity assessments, in what form are sugars presented for assessment?

Birch: Analysts are presented with three sucrose solutions graded in concentration and a further test solution and are asked to rank them in order of sweetness. The relative sweetness was determined on a molar basis.

Parker: You mention that D-viboquercitol is very sweet; how does this compare with sucrose?

Birch: At least as sweet as sucrose, possibly twice as sweet.

Taste Panels and the Measurement of Sweetness

H. W. SPENCER

Lyons Central Laboratories, London, England

ABSTRACT

To a flavour panellist the word 'sweetness' means perception of the basic 'sweet' taste. Although mainly gustatory, this perception also involves response to texture and olfaction (for example, the 'sweet' smell of a rose). Valid assessment of sweetness requires the use of selected and trained judges under controlled test conditions.

The use of laboratory panels of trained judges, by means of standard test procedures, is described: (a) *to determine equivalent sweetness of dextrose and sucrose solutions,* (b) *to compare sweetness in acidified sucrose solutions,* (c) *to determine equivalent sweetness of acidified sucrose and acidified invert sugar* (*dextrose and fructose*) *solutions, and* (d) *to assess the sweetness of acidified solutions of mixed sweeteners by difference testing and to compare these results with those obtained by using the flavour profile technique.*

The use of taste panels by other workers is discussed, including the following: (a) *Woskow—the effect of a flavour potentiator on sweetness,* (b) *Stone and Oliver—the synergistic effect of mixtures of sugars and sweeteners, and* (c) *Yamaguchi* et al.*—using a large taste panel to measure relative sweetness.*

INTRODUCTION

The basic sweet taste is perceived mainly by the gustatory or taste sense though texture and olfaction may also be involved. In this paper only the sense of taste will be considered.

To a flavour panellist measurement of sweetness means estimating the degree of intensity of the sweet sensation that is perceived on the front area of the tongue when sampling a food.

There is much confusion about the way in which the perception of sweetness is tested. The only reliable procedure is by means of a trained panel of judges and I shall outline the manner by which this panel can be selected and trained to perform as a human instrument. Some examples of its use in a food laboratory will be given.

CONDITIONS OF TESTING

There are a number of factors which can affect the perception of sweetness. It is therefore essential to standardise the conditions of testing. These are some of the contributing variables.

Temperature

It was found[1] that D-fructose was less sweet at higher than at lower temperatures when compared with sucrose. Therefore assessments should always be made at the same temperature.

Concentration of the Sweetening Substance

The sweetness of different sugars relative to sucrose varies with the concentration. For example, glucose at a low level is only about half as sweet as sucrose but at a high concentration it has almost the same sweetness as sucrose.

The Medium

Sixty-five per cent sugar syrup tastes much sweeter than 100% 'dry' sugar.

The Age of the Material

Freshly prepared glucose tastes sweeter than a solution made a few hours earlier because the less sweet β-D-isomer is formed during this storage period.

Fatigue of the Judges

Both psychological strain and an excessive number of samples should be avoided.

In standardising the conditions of testing it is recommended that the flavour panel room should be maintained at 21°C and illuminated by constant natural daylight lighting with the facility to switch to coloured lighting (for example, by filters) in order to disguise any difference in the appearance of the samples being compared. The conditions of testing have been discussed in greater detail by the author.[2]

SELECTION AND TRAINING

Selection

Whichever kind of sensory analysis of sweetness is being considered, potential judges should be selected by means of tests of their sensitivity.

Test 1

It is essential that all panel judges can distinguish the basic tastes without error. Sweetness is not usually confused with any other basic taste, but it should be remembered that normally judges have to assess sweetness in mixtures of the basic tastes, some of which are confused by something like 40% of the population. For this first test duplicated samples representing suprathreshold levels of the four basic tastes are recommended (such as 0·07% citric acid, 2% sucrose, 0·2% sodium chloride and 0·07% caffeine) in mains drinking water.

Test 2

Those persons passing the first test should then show their ability to rank different concentrations of sucrose in order of sweetness (such as 7·5, 10, 12·5, 15% in mains water). In the author's laboratories, the potential judges resulting from these two tests were then given a selection of odorants to describe and/or identify. This was because the ultimate flavour panel would be required to work on flavour profiles and other flavour problems requiring sensitivity to a variety of flavourants. However, this is probably not essential to a group of judges concerned only with assessing sweetness. It is therefore not included in the selection procedure here.

Training

This should consist of the test samples that are simultaneously presented to experienced judges in difference testing. About 40 judgements, obtained by considering pairs of samples significantly distinguished by trained judges, should be collected before deciding upon the acceptability of a person for flavour panel work. The number of 'wrong' decisions (established by trained judges' results) is plotted against the total number of judgements. The limit of acceptance or rejection may be set at the 5% probability level and the resultant graph should clearly indicate into which area the potential judge fits. During the training period it is also recommended that potential judges should taste sugar-acid mixtures of known composition to assess both 'sweet' and 'acid' sensations.

ASSESSING EQUIVALENT SWEETNESS: SUCROSE AND DEXTROSE

Two principal methods have been used: (a) threshold determinations and (b) successive comparison of suprathreshold concentrations, although many variations of these methods have been employed.

A concentration of one compound was presented[3] as the standard to 20–30 persons and the two extreme levels of a comparison substance that all persons agreed to be either less sweet or sweeter than the standard were determined. The range of these concentrations was then divided into six or eight different levels and each level compared with the standard.

Biester *et al.*[4] used a 'drop' technique with 30 judges. The mouth was rinsed with distilled water, the tip of the tongue wiped dry with a cotton swab and a drop of the test solution was applied to the surface of the tongue. This curious technique was used to determine the threshold, specified as the lowest concentration of a substance detected as sweet by all judges. This was obviously not a sufficiently standardised technique; for example, the volume of the droplets presented must have varied considerably.

Cameron[5–8] used up to 20 judges who were semi-trained in that if any gave inconsistent judgements they were excluded from the series. Judges were selected by a kind of triangular test. They received three sucrose solutions, either 10, 8·5, 10% or 8·5, 10, 8·5% and knew that the first and last samples were of equal intensity. They assessed whether the middle sample was stronger or weaker than the others; no errors were allowed. About 40% of the candidates were successful, a similar result to that obtained when selecting judges by using the basic tastes in the author's laboratories.

In Cameron's method of assessing relative sweetness, developed from a previous method,[9] three solutions were presented. Judges knew that the samples consisted of two concentrations of one sugar and one concentration of another and they assessed whether the different sugar was less, equally or more sweet than either concentration of the other sugar. It is considered that the paired comparison test would have been a simpler form of presentation, where judges would have selected the sweeter sample and the significance of correct selections would have provided the same information.

Schutz and Pilgrim[10] used a memory-stimulus technique, where samples were scored on a nine-point 'nil' to 'extreme' scale of sweetness. In each test, 10 of 15 untrained judges assessed randomised

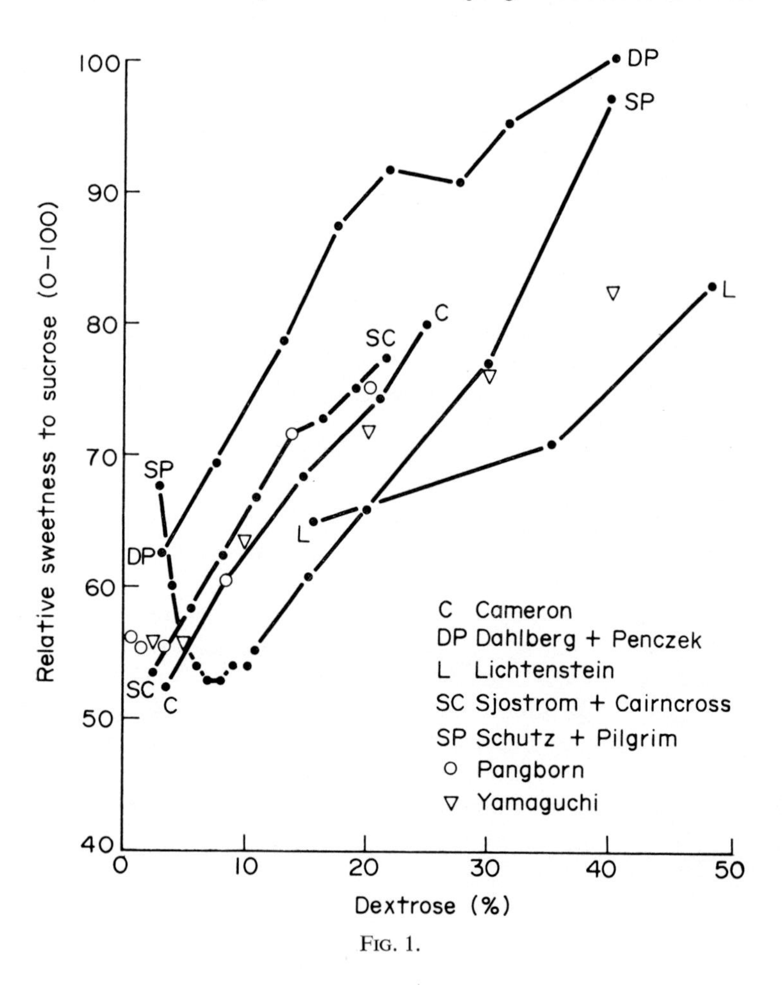

FIG. 1.

duplicated 6 ml samples of a five-step geometric series without swallowing, *i.e.* 10 samples were assessed one at a time, rather a lot for untrained judges. In the case of sucrose they received 2, 4, 8, 16, 32 % w/v solutions. It is considered that a nine-point scale is very wide especially as the judges were not trained. This was a memory technique since judges had no declared standard against which to compare the sweetness.

After 15 daily tests, Pangborn[11] selected from 26 persons, 12 who could distinguish small sucrose sweetness differences and used the paired comparison test to establish the equivalent sweetness of dextrose solutions to a range of 0·5 to 15 % sucrose solutions.

These studies have been reviewed[12] and the author has summarised them including more recent results[11] (Fig. 1).

In Fig. 1, the concentration of dextrose is plotted against the relative sweetness value when sucrose has a value of 100. The graphical diagram shows that Dahlberg and Penczek[9] obtained higher values that the others, possibly because they used only three to five judges. Schutz and Pilgrim's curve was U-shaped,[10] but their memory-scoring technique was unique. Lichtenstein[13] obtained lower results at higher concentrations of dextrose but he used a modification of Biester *et al.*[4] 'drop' technique which was criticised earlier. Sjoström and Cairncross[14] and Cameron[8] obtained similar results and their findings were confirmed more recently by Pangborn.[11]

Mixtures of Sucrose and Dextrose

It has been claimed[9] that a 'supplemental' sweetness action exists when one sugar is added to another. This was confirmed by Cameron, but he noted that there was no apparent enhancement if the results were calculated in terms of dextrose, when a simple addition was found. These results are summarised in Table 1.

The relative sweetness of dextrose in mixtures with sucrose is higher than when it is assessed in pure solutions[12] (compare Table 1 and Fig. 1). When one-fifth to one-half of the total sucrose is replaced, the dextrose relative sweetness is 91–102; when a greater proportion of the total sucrose (two-thirds) is replaced, the dextrose relative sweetness is reduced to 85. According to Fig. 1, a concentration of at least 30–40 % dextrose is needed to obtain a relative sweetness of about 90 in dextrose alone, compared with only 5–10 % dextrose in mixtures containing sucrose.

TABLE 1

Sucrose–dextrose mixtures

Equi-sweet sucrose solution level	Sucrose level (S)		Dextrose level (d)	Proportion of S replaced by d	Total sweetness as sucrose	Total sweetness as dextrose	Relative sweetness of dextrose
15% (C)	5%	+	11·8%	$\frac{2}{3}$	15 = 12·6	20·5 = 20·1	85
15% (C)	10%	+	5·5%	$\frac{1}{5}$	15 = 13	20·5 = 20·2	91
15% (D + P)	10%	+	5·3%	$\frac{1}{3}$	15 = 13·5	—	94
20% (C)	10%	+	10·2%	$\frac{1}{2}$	20 = 16·4	25·0 = 24·9	98
25% (D + P)	16·7%	+	8·3%	$\frac{1}{3}$	25 = 22·7	—	100
40% (D + P)	26·7%	+	13·0%	$\frac{1}{3}$	40 = 37·1	—	102

C = Cameron, D + P = Dahlberg + Penczek.

TABLE 2

Sweetness of acidified sucrose solutions

Citric acid (anhyd. %wv)	Judges																Total correct judgements	SL
	1	2	3	4	5	6	7	8	9	10	11	12	13	14	15	16		
0·01	1	2	0	2	3	2	1	0	2	2	2	2	1	2	2	1	25/48	NS
0·02	0	1	1	2	1	1	2	2	1	1	2	2	3	0	0	1	20/48	NS
0·04	3	3	3	2	2	3	3	1	3	3	2	3	3	1	0	2	37/48	VS
0·06	3	3	3	2	3	2	3	0	2	1	3	3	3	2	2	3	38/48	VS

Sweetness of Acidified Sucrose Solutions

In the author's laboratories, paired sucrose solutions were presented in triplicate to 16 judges in two separate sets. Citric acid (0·01, 0·02, 0·04 or 0·06%) had been added to one solution of each pair and the task was to select the sweeter of the two solutions, *i.e.* this was a paired comparison test. Judges received six pairs of randomised coded solutions in each set. Rinsing of the palate with mains drinking water, as used to prepare the solutions, was encouraged between pairs to lessen palate fatigue, but not within pairs to avoid confusion of the taste impression of a pair of samples. The results are summarised in Table 2.

Acidification by 0·04 or 0·06% citric acid significantly reduced the sweetness of the sucrose solutions. Lower acid levels had no significant effect on the sweetness. Much more highly trained judges found that the sweetness of 2% sucrose solution was depressed by the addition of as little as 0·01% citric acid.[15]

Equivalent Sweetness of Acidified Sucrose and Acidified Invert Sugar Solutions

In these experiments in the author's laboratories, the standard mixture consisted of 10% w/v sucrose and 0·3% w/v citric acid in mains drinking water. The paired comparison technique was followed. Seven experienced judges assessed four pairs of samples, each pair being one standard and one comparison invert sugar solution in random order. Swallowing was not allowed and soda water was used to rinse the palate between pairs. Two comparison tests were carried out; in one, the comparison samples contained 11% invert sugar and in the other 12·5% invert sugar, both acidified to the same level as the standards. Judges were asked to select the sweeter sample. These are two-tailed tests because either the standard or the test samples may be selected as the sweeter. The results are summarised in Table 3.

The significance level was determined by reference to published tables. From these results it was concluded that the 10% sucrose standard was just significantly sweeter than 11% invert sugar, and 12·5% invert sugar solution was just significantly sweeter than the standard. The sweetness level of acidified invert sugar equivalent to that of acidified 10% sucrose was thus between 11·0 and 12·5%.

TABLE 3

Sweetness of acidified sugar solutions (%w/v)

Sugar level	Relative sweetness	Corrected sucrose equivalent	Number of selections	Significance level
10·0 sucrose	10·5	10·3	20	JS
11·0 invert sugar	11·0	10·8	8	NS
10·0 sucrose	as before		8	NS
12·5 invert sugar	12·2	12·0	20	JS

NS = not significant; JS = just significant (5% probability level)

Sweetness of Acidified Solutions of Mixed Sweeteners

Paired comparison tests

Again using judges in the author's laboratories, four different mixtures of sucrose, sodium cyclamate, saccharin, dextrose monohydrate and citric monohydrate were assessed in random pairs using the conditions of presentation and testing already described. The composition of the samples is given in Table 4. The results obtained from the judges are summarised in Table 5.

TABLE 4

Composition of mixtures (%w/v)

	A	B	C	D
Sucrose	4·6	4·6	2·6	1·9
Na cyclamate	0·02	0·02	NIL	0·04
Saccharin	0·005	0·005	0·008	0·003
Dextrose monohydrate	NIL	NIL	1·7	1·8
Citric acid monohydrate	0·2	0·3	0·2	0·2
Relative sweetness	4·1	4·1	3·9	4·0

Solution B was found to taste less sweet than A presumably because it contained more acid, all the sweet ingredients being at the same level. Solution A was sweeter than C probably because of its higher sucrose content and the presence of sodium cyclamate, although C contained 50% more saccharin and some dextrose

TABLE 5

Paired comparison tests

Comparison samples	Judges (a)	(b)	(c)	(d)	(e)	(f)	(g)	Total selections*	Significance level	Conclusion
A v. B	4	4	3	4	3	3	4	25/28	VS	A > B
A v. C	4	4	4	4	4	4	2	26/28	VS	A > C
B v. C	4	0	4	0	3	3	1	15/28	NS	B = C
A v. D	4	4	2	2	0	3	3	18/28	NS	A = D

VS = very significant (0·1% probability); NS = not significant.
* Selections in the same direction.

monohydrate that was not present in solution A. In addition, although of similar acidity, A had a higher relative sweetness than C.

Comparison of paired tests and the flavour profile technique

During a number of profile sessions, the intensity of 'sweet' and 'acid' was determined by the same judges. Their assessments are summarised in Table 6.

TABLE 6

Taste profiles of solutions

Flavour characteristic	A	B	C	D
Sweet	2–3	1–2	1–2	2–3
Acid	1–2	2–3	x–1	1–2

Intensity scale: x = threshold, 1 = slight, 2 = moderate, 3 = strong.

The sweetness assessment was completely correlated with the paired test results shown in Table 5. It is recommended, when possible, to confirm sensory results by an alternative method such as that illustrated here before attempting to correlate with chemical analysis.

SYNERGISM

More recent work[16] helps to relate the sweetness assessments of the mixtures just considered to their composition. Synergism may be said to occur when the total effect of the stimulus is greater than the sum of the contributions of the individual components of a mixture. Steven's method of magnitude estimation was used. Six highly trained judges received a solution stated to have a sweetness intensity of '10' and a random series of samples of intensity above and below this. Each sample was scored as a factor of '10' and a blind reference standard was also included at the geometric mean. The series consisted firstly of individual sugars or sweetening agents, then mixtures. The mixture experiments showed synergistic effects between

(*a*) sucrose and dextrose—about 20 to 25% synergism,

(*b*) sucrose and fructose—very slight, probably because these two sugars have similar sweetening properties,
(*c*) dextrose and calcium cyclamate—some synergism.

Saccharin mixtures showed an additive, not a synergistic, effect.

The confirmation of synergism between dextrose and calcium cyclamate and between sucrose and dextrose could explain why solution D in the previous section, containing all four substances was as sweet as solution A which had about two and a half times as much sucrose, half as much cyclamate and no dextrose (Table 4). The large amount of synergism between sucrose and dextrose—(*a*) above—probably also explains why solution C tasted as sweet as solution B; being more acid, B would be expected to taste less sweet, but the synergistic effect of sucrose and dextrose probably increases the apparent sweetness in C (there was no dextrose in B) just sufficiently to make them seem equally sweet. The same sucrose—dextrose synergistic effect probably explains why solutions A and D were equi-sweet as outlined in Table 7.

The magnitude (intensity) estimation technique[16] is an improvement upon the memory technique[10] since it does include a standard level to anchor the judgements. However, unless extremely well-trained judges are used it is considered to be far less accurate than the paired comparison technique since a range of different concentrations is estimated during the same test.

TABLE 7

Sugar content of solutions A, D (see also Table 4)

	A	D	Synergistic effect	Resultant sweetness of D
Sucrose (%)	4·6	1·9		
Dextrose (%)	NIL	1·8 = 1·35 sucrose		
Total sugar (%) (as sucrose)	4·6	3·25	25%	4·06

FLAVOUR POTENTIATION OF SWEETNESS

A flavour potentiator acts as a sort of catalyst to augment the perceived flavour without introducing an additional flavour. Paired comparison tests have been used[17] to illustrate the effect of concentrations of a 50:50 mixture of two 5′-ribonucleotides (inosinic acid and guanylic acid) on the perception of the basic tastes. In the case of sweetness, the standard was 5% sucrose and the results are shown in Table 8 as the number of selections of the sweeter sample.

TABLE 8

Sucrose and ribonucleotides

Comparison sample (Amount of R added to 5% sucrose)	Number of selections of comparison sample*	SL
0·004%	46/90	NS
0·008%	55/90	JS
0·012%	53/90	NS
0·016%	61/90	VS
0·020%	59/90	S

R = 5′-ribonucleotides. * Transposed from stated percentage selections SL = significance level, NS = not significant, JS just significant (5% probability level) S = significant (1% probability level), VS = very significant (0·1% probability level).

In the case of the result for 0·012% only, two more selections were needed to indicate this concentration to be a just significant level of additive. It is evident from the results that the potentiator increased the apparent sweetness of 5% sucrose solution.

USE OF A LARGE TASTE PANEL

Japanese workers[18] used the paired comparison technique. They selected 100 most taste-sensitive persons from about a thousand candidates. These persons were then presented with five pairs of solutions at a time, each pair consisting of one sucrose and one other sugar/sweetening agent solution. The equivalent sweetness was determined according to the number of judgements selecting the sucrose as the sweeter sample. Either 25 or 50 total judgements were obtained for each comparison.

It is difficult to see the need for such a large panel of judges (100 persons) when only 25 or 50 judgements were actually obtained for each substance. Pangborn and the author obtained about

TABLE 9

Equivalent sweetness to sucrose (%wv)

Sucrose equi-sweet level	Glucose	Fructose	Sodium cyclamate	Sodium saccharin
1·24				0·0025
1·40	2·5			
1·62			0·05	
2·13				0·005
2·80	5·0			
3·23			0·10	
3·3		2·5		
3·54				0·010
5·41				0·020
5·84			0·20	
6·32	10·0			
6·66		5·0		
7·40				0·040
9·18				0·080
12·30		10·0		
14·40	20·0			
23·00	30·0			
24·70		20·0		
33·00	40·0			

30 judgements, three or four from seven to ten judges. This considerably reduced the inter-person variation. Although selected, the Japanese judges did not constitute a trained panel. In effect, the judges were sensitive consumers.

Some of their results are shown in Table 9.

By converting the Japanese glucose values to relative sweetness (sucrose = 100) and comparing with Fig. 1 in the Sucrose and Dextrose section of this paper, it is seen that the results for 2·5 to 20% glucose were similar to the findings of Cameron and Pangborn.

Interactions of ingredients of mixtures of sweet substances were also studied by the same Japanese workers.[18] Unlike previous workers, they restricted the use of the term 'synergism' to reactions where the resultant sweetness was greater than the sum of the sweetnesses of the ingredients *whichever of the ingredients was used as the reference standard for calculating the overall sweetness.* They therefore excluded mixtures of sucrose and glucose because when the overall sweetness is calculated from glucose it is additive. They showed that their specifically defined synergism occurred between (i) sucrose and fructose (as stated by Stone and Oliver[16]), (ii) sucrose/fructose and sodium cyclamate, (iii) fructose and sodium saccharin, and (iv) sodium cyclamate and sodium saccharin.

REFERENCES

1. Tsuzuki, Y. and Yumazaki, J. (1953). *J. Chem. Soc. Japan, Pure Chem. Sect.*, **74,** p. 596.
2. Spencer, H. W. (1971). *The Flav. Ind.* (in press).
3. Paul, T. (1922). *Zeitschr. Untersuch. Nahr. u Genussm.*, **43,** p. 137.
4. Biester, A., Wood, M. and Wahlin, C. (1925). *Am. J. Physiol.*, **73,** p. 387.
5. Cameron, A. T. (1943). *Trans. Roy. Soc. Can.*, Ser. 3 (Sect. V), **37,** p. 11.
6. Cameron, A. T. (1944). *Can. J. Res.*, **22** (Sect. E), 3, p. 45.
7. Cameron, A. T. (1945). *Can. J. Res.*, **23** (Sect. E), 5, p. 139.
8. Cameron, A. T. (1947). *Sugar Res. Found., Inc.*, Sci. Rep. Ser. No. 9, N.Y.
9. Dahlberg, A. C. and Penczek, E. S. (1941). *N.Y. State Agric. Exp. Sta., Tech. Bull.*, p. 258.
10. Schutz, H. G. and Pilgrim, F. J. (1957). *Fd. Res.*, **22,** p. 206.
11. Pangborn, R. M. (1963). *J. Fd. Sci.*, **28,** p. 726.
12. Niemann, C. (1960). *Zücker u. Süsswaren—Wirtschaft*, **13,** pp. 10–14, 620, 629–630, 674, 677–8, 706–7, 756–760.
13. Lichtenstein, P. E. (1948). *J. Exp. Psychol.*, **38,** p. 578.

14. Sjoström, L. B. and Cairncross, S. E. (1955). *Adv. Chem. Ser.*, **12,** p. 108.
15. Pangborn, R. M. (1961). *J. Fd. Sci.*, **26,** p. 648.
16. Stone, H. and Oliver, S. M. (1969). *J. Fd. Sci.*, **34,** p. 215.
17. Woskow, M. H. (1969). *Fd. Techn.*, **23,** pp. 32, 37.
18. Yamaguchi, S., Yoshikawa, T., Ikeda, S. and Ninomiya, T. (1970). *Agr. Biol. Chem.*, **34,** 2, p. 181.

DISCUSSION

Macleod: 1. Do you supply your panellists with a constant volume of solution for tasting, since at low concentrations it will be the number of molecules of the compound entering the mouth that is important rather than simply the concentration of the solution?

2. It is generally accepted that a stimulating solution held motionless in the mouth elicits no response, and movement of the tongue or swirling the solution around the mouth is necessary to receive the appropriate stimulus. Do you give your panellists any instructions with regard to this factor?

3. Does whether your panellists are smokers or non-smokers affect their sensitivity to sweetness?

4. Do you have any opinions on whether there are four or six basic taste sensations and the generally accepted theory that certain areas of the tongue are particularly sensitive to one taste stimulus (*e.g.*, the tip of the tongue sensitive to sweetness)?

Spencer: 1 and 2. Regarding how much to give the judges: in general, in assessing one sample, 20 ml or sometimes 10 ml are taken into the mouth, swirled around the tongue, and spat out. If you have a pair of samples, you taste the first sample and then go on to the second sample. You are allowed to go back to the first sample, and then rinse before going on to the next pair. It is difficult. You cannot give too rigid instruction on this and you have to play it along. Once you have tasted, the main thing is to stick to these conditions for the rest of the experiment.

3. We have on our panel people who are smokers and non-smokers, about 50:50. The sensitivity is not affected by the normal amount of smoking, but heavy smoking decreases sensitivity.

4. More and more substances are being discovered which do seem to have mixed reactions on the tongue, but I think from the point of view of selecting judges, you can help by familiarising them with the four basic tastes and, if they are having difficulty between sour and bitter, they will register bitterness at the back of the tongue and sweetness at the front of the tongue.

Dixon: Have you any experience of the relationship between viscosity and sweetness detection, as it has been reported in the literature that more viscous materials possess greater relative sweetness compared with sucrose?

Spencer: I have not looked into this problem very much as I have been mainly concerned with drinks where texture does not play a great part, but

I am sure that viscosity plays an important part in the perception of sweetness.

Coulson: As a follow up to Miss Dixon's question on the influence of viscosity perhaps, Mr Chairman, through you and the speaker, I could put a question to Prof. Pangborn. You refer in your book (Amerine, Pangborn and Roessler) to the influence of macromolecules on sweetness. Do you have further information on the effect, leaving aside that of viscosity, on sweetness of say carboxymethyl cellulose starch or gelatin—in perhaps equimolar concentrations.

Pangborn: We have recently completed an investigation on the interaction of five food gums with sucrose, sodium saccharin, sodium chloride, citric acid and caffeine (Pangborn, R. M., Trabue, I. M. and Szczesniak, A. S. (1972), *J. Texture Studies*, in the press). In general, the tastes of sodium chloride and saccharin were enhanced, whereas sucrose, citric acid and caffeine were reduced by the addition of gums. A study is in progress on the interactive effects of the same gums with the aroma of four distinct compounds, acetaldehyde, acetophenone, butyric acid and methyl sulphide.

O'Mahony: Are group-interaction effects allowed for in a taste panel? I am thinking of how to guard against drift-effects where group interaction may cause decisions to be more extreme than individual decisions.

Spencer: This obviously can influence results and one has to guard against it. In the case of difference tests, you are working in isolation, so that your decisions are not influenced by your neighbour. As for group interaction, it is most important on a flavour panel to provide standards which all agree upon. Some people are more sensitive than others to certain flavours. You need people who work in somewhat similar fashion and do not have too large a range of difference between them.

Adams: 1. Have you any information on surface tension effects?

2. With your taste panels, do you use any of the substitute ureas or tannins to obtain a general picture of the panel for comparison with outside panels?

3. Do you have a special panel vocabulary containing descriptive terms suitable for different substances?

Spencer: 1. No experience.

2. These have not been used.

3. A vocabulary is certainly built up to apply to the particular food you are dealing with. It is confined to that food and built up over a number of sessions. The terms are then agreed within the group of judges and only these terms used.

Macdonald: Is there much variation in taste threshold within an individual?

Spencer: For sweetness it is about 0·3/0·4% sucrose. We have had limited experience in this area, but, of those tested, there was very little day-to-day variation.

Cunliffe: Comments made during the Symposium on the relative sweetness of sugars and other sweeteners relate to sugar solutions at relatively low concentrations. How reliable are panel tests conducted on products such as jams, jellies and other food products containing much higher levels of sugar?

Spencer: This comes back to the consistency of the panel. What you really want are people who are similarly sensitive, but not necessarily with low sweetness thresholds. Unless the panel is consistent, you cannot use them.
Weaver: Does the speaker use mains or distilled water when doing threshold and basic taste reaction tests?
Spencer: We are not happy with distilled water because you get metallic tastes which are very variable. The variation in mains water is not great and we always use it. I find it useful to keep a sample of mains water, something like 10 litres, overnight in a large vessel to allow the chlorine to evaporate.
Swindells: Following the question on variation of threshold from day-to-day, has Mr Spencer encountered much variation in threshold, or discrimination in sweetness, during the course of a single day? For example it is my impression that discrimination in soft drink flavours is less during the afternoon than before lunch.
Spencer: The time of testing is limited in that you have to allow for meal times, and in the main, having decided on a time of testing, I think you should stick to it. If you test a lemon drink in the afternoon, then always test it in the afternoon. Try to keep as consistent as you can in your conditions of testing.

It is a bit difficult to make sweeping statements on this because much is dependent upon the group of people. We found that, in a series of experiments with lemon drinks, our people were more sensitive in the afternoon.

The Industrial Potential of Cereal-Based Sweeteners

K. SELBY and J. TAGGART

The Lord Rank Research Centre,
Lincoln Road, High Wycombe, Bucks., England

ABSTRACT

The replacement of sucrose by cereal-based sugars in food products is reviewed with particular reference to sugar confectionery, flour confectionery, jams and marmalades, soft drinks and beer. The development of low-calorie sucrose substitutes is described. The effect of Britain's entry into the European Economic Community on the economics of cereal-based sweeteners is discussed.

INTRODUCTION

For many years, the food industry has used sucrose in very large quantities, principally as a sweetener, but also because of its valuable effect as a preservative and to improve the appearance of many food products. In the last year for which detailed figures are available, the usage of sucrose was over 2½ million tons. An analysis of sucrose disposals is shown in Table 1.

There is a good deal of evidence to support the view that sucrose is a primary factor in producing dental caries,[2] especially in children. There is some evidence to suggest that sucrose may be implicated in the aetiology of some heart diseases,[3] and, taking a broad view, that high sucrose intake may be bad for our general health. If these views are substantiated by further research, an impetus will be given to studies of methods of reducing the direct and indirect consumption of sucrose; there is already a body of opinion that suggests, on public health grounds, that these measures should be introduced now.[4] To anticipate this trend, several major food manufacturers have commissioned research programmes to develop replacements for

TABLE 1

Disposal of refined sugar for food in the UK.
Mean of 1965 *and* 1966[1]

	Tons × 10^3
Chocolate and sugar confectionery	359
Cakes and biscuits	175
Jams and marmalades	137
Soft drinks	130
Beer and British wines	80
Syrup and treacle	53
Canned goods (fruit, vegetables, soups)	44
Table jellies	33
Condensed milk	30
Ice cream	18
Cereal breakfast foods	15
Pickles and sauces	14
Cider	12
Cake and bun mixtures	5
Canned puddings	5
Quick-frozen foods	1
	1111
Direct consumption	1537
Total	2648

sucrose. Since the total market is some $2\frac{1}{2}$ million tons, it would seem logical that the starch-bearing cereals might provide part, if not all, of the replacement. It must be stressed that the bulk of the development is not being carried out solely on public health grounds; there is a sound economic basis for the use of cereal sugars, which is discussed at the end of this contribution.

In reviewing the industrial use of sucrose and glucose syrups one finds that the sweetness of sucrose is not as important as might be thought. Table 1 shows that there are four major industrial outlets for sucrose, which together account for almost 75% of the indirect consumption. Obviously, these areas would provide the richest return for the cereal sugar manufacturers.

CHOCOLATE AND SUGAR CONFECTIONERY

As Table 1 shows, this is a very large market, accounting for over 30% of the indirectly-consumed sucrose in the UK. Cereal sugars have been used for some time in products of this type.[5] Ranks Hovis McDougall (RHM) have been working closely with a major school of preventive dentistry and more recently with the British Food Manufacturing Industries Research Association to develop sucrose-free sweets; at the moment small-scale placement trials are under way to assess the acceptability of these materials by children. Because of the unusual properties of wheat sugars,[6] it has proved surprisingly easy to prepare certain types of sugar confectionery, particularly caramels, fondant creams with and without coconut, nougat and butterscotch. Some difficulty is experienced in preparing the 'crystalline' type of sweet and jelly where the physical properties of sucrose are clearly very important. However, it seems likely that the remaining technical problems can be overcome. As previously reported[6] one of the Group's wheat-based products, Trudex, has the ability to form a semi-crystalline solid sweet-like product containing half its own weight of oils such as cod-liver oil or liquid paraffin. It is anticipated that this material could be used as a carrier for fat-soluble essential compounds and although it is still in the early development stage this outlet looks promising.

CAKES AND BISCUITS

Not unnaturally, a group like RHM that has a commitment both to wheat sugars and to flour confectionery is interested in combining the two. Based on developments in the USA, the replacement of sucrose by high-maltose products derived from wheat starch was examined. In some cases, the lower level of sweetness of the cereal-based product was a problem. The specification for a synthetic sweetener to overcome this problem would be fairly difficult to meet; for example, the compound would have to withstand prolonged exposure at a fairly high temperature and have a substantial shelf-life in solution. This problem will probably be solved by the partial replacement of, say, up to 30% of the sucrose used, and the applications will be principally in fruit cakes like Dundee and Madeira.[7] The partial replacement of sucrose by low DE wheat sugar in pastries has led to the development of pie and flan casings with improved resistance to moisture migration.[6] Although not related to sweetness, this phenomenon, which as yet is only partially understood, is an important product-plus. Developments in the biscuit field are at an early stage and do not merit discussion here.

JAMS AND MARMALADES

The main function of sugars in jams and similar products is to preserve the fruit by generating a high osmotic pressure and thus to inhibit the development of microbial spoilage. For some time, the industry has used a partial replacement of sucrose by glucose syrup to prevent the crystallisation of sucrose and to enhance the flavour of the product, since the palate may become saturated by too much sweetness.[8] One of our major subsidiaries, Energen Products Limited, has recently introduced a range of low-calorie jams in which a synthetic sweetener is used to give a characteristic 'jam' taste, but in which the level of sucrose and accordingly the calorific value has been substantially reduced. The formulation of low-sucrose jams is therefore in hand. The problems are similar to those relating to fruit-pie fillings in which it was expected that any organoleptic disadvantages encountered in the use of synthetic sweeteners could be minimised by the incorporation of more strongly-flavoured fruit pulps. It has been found in practice, however, that because of the lower relative sweetness of the glucose-containing cereal sugars, the natural flavour of the fruit is greatly enhanced, and often becomes preferable to the comparable sucrose-containing product.

SOFT DRINKS

Again based on experience in the USA, sucrose-free low-calorie drinks containing synthetic sweeteners have been developed and are being marketed in the UK.[9] A typical lemonade, for example, contains about 4 to 5% sucrose, but recently in the USA fructose/gluconic acid mixtures have been introduced into this type of product.[10] The high relative sweetness of fructose is utilised with or without a synthetic sweetener such as saccharin; the cost is minimised by the addition of gluconate which helps to mask the bitter after-taste but does not contribute calories. In this country RHM are exploring the use of low DE cereal sugars in soft drinks, and as described later, the possibility of producing fructose from cereal starches is under intensive study. In addition to the applications described above, sucrose is used in the brewing industry, both as an adjunct in the initial fermentation and as a priming sugar.[11] In some beers like milds and stouts a proportion of the sucrose comes through into the final product. Of the 80,000 to 100,000 tons of sucrose used by the industry per annum, between 90% and 95% is fermented, so that between 4000 and 10,000 tons of sucrose per annum will be consumed by beer drinkers. A great deal of effort is being brought to bear on the production of cereal adjuncts and wort replacements for the brewing industry[12]; it seems likely that acceptable beers and ultimately wines and spirits may be produced using very sophisticated syrups based on cereals; from the Group's point of view it is hoped that these will be based principally on wheat. In this field again, difference in sweetness is not a major barrier to the introduction of cereal-based products.

LOW-CALORIE SWEETENERS

The development of dextrin-based sucrose substitutes began in America. One of the first commercially-acceptable low-calorie sweeteners was described in a patent in 1965 by Alberto-Culver Inc.,[13] since when there have been innovations by Monsanto, Pillsbury and Procter and Gamble. In essence, the preparations were made from a low DE cereal syrup containing either cyclamate or saccharin. The sweeteners were used at about 5% w/w for cyclamate with or without about 0·5% saccharin. An Alberto-Culver innovation was the incorporation of gum acacia or gelatin to simulate the crystalline appearance of sucrose. The mixture was simply vacuum-dried, during which a porous cellular structure was induced, and finally comminuted to the appropriate particle size. Three competing products have been launched in the United States under the names 'Sprinkle Sweet' (Pillsbury), 'Scoop' (General Foods) and 'Twin' (Alberto-Culver). The RHM product in this field is known as Energen Sweetness, and contains only saccharin as the sweetening agent. To a certain extent, the use of low DE wheat sugar ameliorates the bitter after-taste of saccharin. It has been reported that the after-taste can be removed completely by the addition of 0·5 ppm of a 5′-purine-nucleotide.[14] The low bulk density of these products makes them easy to pack and distribute, and their major selling point is that, because of their low bulk density, spoon for spoon they produce the same intensity of sweetness as sucrose, but at a lower calorific value. The size of the market for a product of this type is always difficult to estimate because of seasonal factors, and re-formulation exercises. The product made by RHM is aimed at an outlet of some 7000 to 10,000 tons per annum, representing about 1% of the direct sucrose consumption. Figures for consumption in the USA are difficult to obtain since the highly competitive atmosphere makes companies reluctant to release their true sales figures. The market is certainly large and from the size of the companies that have entered it one may deduce that it is a rapid-growth area.

SUMMARY

The methods of production of the various cereal sugars described here, have been comprehensively reviewed.[6,15] In simple terms, starch is isolated from the appropriate cereal and hydrolysed using acid and/or a variety of microbial enzymes to produce the desired degree of depolymerisation. The refined final products are normally sold as syrups, but a variety of solid glucose products are available. Future developments in this field are likely to include the introduction of insolubilised enzymes[16] which will allow the development of a continuous process, and more significantly from the point of view of sweetness the conversion of glucose to invert sugar using glucose isomerase.[17] This latter development should give a low-cost cereal sugar with the sweetness of sucrose and should open up a large area of the market.

The prospect of Britain's entry into the European Economic Community will affect the long-term economics of cereal sugar production. It seems likely that the price of sucrose will rise from its present value of around £70 per ton to about £90 per ton. Because of an EEC regulation governing the price of cereals for the starch industry,[18] the raw-material cost of wheat-based processes will almost certainly remain constant, and will probably fall. This will give added impetus to the development of wheat-based sucrose replacements.

The industrial potential of cereal-based, and particularly wheat-based, sweeteners is very great. The partial replacement of indirectly-consumed sucrose has presented no major difficulty, and the future trends of the industry will facilitate the development of a realistic alternative to sucrose. In the direct-replacement field, there are still many problems to be overcome; in this context the advances that are being made in the fields of intense sweeteners of natural origin and the chemical basis of sweetness, described in this volume, are particularly welcome. There is some evidence that public health considerations, and clear evidence that economic factors will favour the increasing trend towards cereal-based sweeteners.

REFERENCES

1. Brook, M. (1971). *Sugar*. Ed. by Yudkin, J. and Edelman, J., Butterworths, London (in the press).
2. *Report of Committee on Medical Aspects of Food Policy, Panel on Cariogenic Foods* (1969). *Br. dent. J.*, **126,** p. 273.
3. Paul, D., MacMillan, A., McKeen, H. and Park, H. (1968). *Lancet*, **2,** p. 1049.
4. Hitchin, D. A. (1970). *Proceedings of the International Health Conference, Edinburgh*, p. 29, Royal Society of Health, London.
5. Maiden, A. W. (1970). In *Glucose Syrups and Related Carbohydrates*, p. 3, Ed. by Birch, G. G., Green, L. F. and Coulson, C. B., Elsevier, London.
6. Selby, K. and Wookey, N. (1970). *ibid.*, p. 46.
7. Pannel, R. J. H. (1968). *Br. Baker*, **157,** p. 68.
8. Palmer, T. J. (1970). *Proc. Biochem.*, **5**(5), p. 23.
9. Brook, M. (1969). *R. Soc. Hlth J.*, **89,** p. 140.
10. Dawes Laboratories Inc. (1970). *Food Processing*, **31**(12), p. 23.
11. Parker, K. J. (1970). In *Glucose Syrups and Related Carbohydrates*, p. 81, Ed. by Birch, G. G., Green, L. F. and Coulson, C. B., Elsevier, London.
12. Maule, A. P. and Greenshields, R. N. (1970). *Proc. Biochem.*, **5**(2), p. 39.
13. Norse Chemical Company assigned to Alberto-Culver Inc. (1965). US Patent No. 3,325,296.
14. Takeda Chemical Company (1966). British Patent No. 1,146,446.
15. Knight, J. W. (1969). *The Starch Industry*, Pergamon Press, Oxford.
16. Barker, S. A. and Epton, R. (1970). *Proc. Biochem.*, **5**(8), p. 14.

17. Agency of Industrial Science and Technology, Japan (1968). British Patent 1,103,394.
18. Swann, P. (1970). *The Economics of the Common Market*, pp. 80–3. Penguin Books Ltd, Harmondsworth, Middlesex, UK.

DISCUSSION

Finch: I should like to follow up, if I may, your point about the economic, as distinct from the health aspect of cereal sweeteners by comparing the relative cost of sucrose and cereal-based sweeteners.
Selby: This is a difficult one. It is because of the difficulty of assessing the cost of sucrose. The problem is that it varies in price according to the use to which you put it and whether it qualifies for a rebate. The cost of cereal solids is £70 per ton but people seem to want to take syrups. This reduces the price by approximately £10 per ton which is, in fact, the drying cost. Low D.E. materials, because of lower throughput due to insufficient demand, cost £110—£115 per ton. To get a realistic costing one must think in terms of plant time per hour. Usage in soft drinks may be as much as 10% sucrose, and confectionery and jams are well above this, but I have had no experience of assessing such products in relative sweetness cost terms.
Finch: The use of enzymes in connection with production of starch hydrolysates is under active research in universities. Are there prospects of quick advances?
Selby: It is quite correct to say that the use of solid enzymes produces viable materials but the greatest advances are needed in the sphere of reactor design, filtration being the main problem.
Palmer: I should like to give a little more colour and perspective to Dr Selby's paper. By far the largest amount of cereal sweeteners within this country are manufactured from maize rather than wheat. A number of the problems associated with wheat sugars, and mentioned by Dr Selby, are due to the inferior quality of those materials in comparison with sucrose, or glucose syrups from maize. I have been somewhat daunted by the information presented at this Symposium by our more academic colleagues. A great deal is known of the relative levels of sweetness as determined under almost clinical conditions. The food manufacturer needs information on the sweetness response in the organoleptic environment of the finished product. Can anyone present make comment or give information on sweetness response in the complex carbohydrate/amino acid environment of foods? The possibilities for synergistic effects seem to become almost unlimited. Or am I thinking in molecular proximity distances that are far too large (*e.g.*, tens or hundreds of Å) rather than the 3Å mentioned by Prof. Shallenberger as vital to elicit sweet/bitter responses?
Selby: All I really want to say is "Yes!" As you probably know, we do have a taste panel at Ranks, and we do take this problem very seriously. We have a panel which is trained to pick out particular flavoured foods, dealing with sweetness amongst many other things.

Spence: 1. Would Dr Selby confirm that he was not intending to imply that starch or starch-based glucoses are of lower calorific value than sucrose?

2. Why attack sucrose as posing a health problem and then propose the isomerisation of glucose to fructose, the half of the sucrose molecule likely to be associated with coronary heart conditions?

Selby: 1. I was obviously not intending to imply that glucose syrups contribute less calories than sucrose on a dry weight basis.

2. Fructose/glucose syrups are being developed because the fructose moiety of sucrose is not *definitely* implicated in heart conditions.

Spence: I think the Beecham Group have shown fructose to deposit fat more readily than glucose.

Green: Fructose, when in combination with glucose as sucrose, has been shown to be associated with fat deposition but when administered in admixture with glucose, its effects appear to be less serious.

Shallenberger: If fructose is implicated this problem obviously becomes an emotional one, since fructose is the major sugar in apples and other everyday foods.

Green: In the context of the addition of saccharin to low DE spray dried glucose syrups for sucrose replacement has the lability of saccharin been overcome by Rank?

Selby: I must stress that this sweetener has been used only as a replacement of table-sucrose and not at all in cooking.

Philp: Are you suggesting that the replacement products are less cariogenic than sucrose?

Selby: Yes.

Grenby: Work has been done on rats feeding solid sucrose and glucose syrups with little difference in cariogenic effect; when solutions of these were fed to the rats, there was a marked reduction of cariogenicity due to glucose syrup.

Cunliffe: Since children eat sweets because they like sweet things, will they eat sweets that are less sweet?

Selby: One must remember that although a reduced sweetness may be more acceptable to adult palates, this may not necessarily be the case with children. My daughter seems to find them preferable. The attraction that sweets have for children is probably due more to the social/psychological overlay, as Dr Watson inferred earlier, than to the sweetness response on consumption. I am not convinced that the real reason why the child wants to buy them is because of the sweetness.

Legislative Aspects of Artificial Sweeteners and Other Food Additives

P. S. ELIAS

Department of Health and Social Security,
London, England

ABSTRACT

After a brief historical introduction the UK legislation related to sweeteners is discussed. The functions of the Food and Drugs Act 1955 *are described. The need for food legislation is discussed in detail in relation to food hygiene, unintentional contaminants, naturally occurring toxins, the irradiation of food and intentional food additives. This leads logically to the subject of safety testing of food additives and the evaluation of test results in terms of risk to man. Then follows a brief description of the advisory machinery, the making of regulations and their enforcement with a note on toxicologically insignificant levels.*

HISTORICAL INTRODUCTION

The legislative aspects involved when considering the subject of sweeteners are part of the general control exercised by HM Government over food standards in respect of composition, hygiene and labelling. The earliest statutory controls in this country were directed not at protecting the public but at the prevention of revenue evasion. Only with the realisation that changes in the quality of food could be affected during preparation for sale, did governmental attitudes change and the Bread Act of 1836 was passed. The Pure Food Law of 1860 was the end-result of a prolonged campaign largely carried on by certain medical personalities rightly opposed to the widespread practices of food adulteration which existed at that time.[1] This was followed by the Sale of Food and Drugs Act of 1875 which was a more satisfactory piece of legislation. As a result, food standard legislation was introduced in Britain for such foodstuffs as milk, butter and other dairy products, confectionery, etc. designed to stop defraudation of the consumer and to prevent such appalling abuses as the use of white lead in confectionery, the contamination of food by arsenic, and the elimination of bacterially decomposed food.

Only much later and not until The Food and Drugs Act of 1938, were Ministers empowered to promulgate regulations for the composition and labelling of food. This Act never came into operation because of the outbreak of the 1939–1945 War, food regulations and control being carried out under the Defence Regulations. Controls were thus largely concerned with rationing and price control although some purely compositional regulations covering foods of comparatively minor importance, such as mustard and fish cakes, were enacted. Some of these are still in existence today.

LEGISLATION RELATED TO SWEETENERS

Of more immediate interest are the only two Orders of 1953 concerned with sweeteners, namely the Food Standards (Saccharin Tablets) Order, 1953 and the Artificial Sweeteners in Food Order, 1953.[2,3] The former revised the standard for saccharin tablets previously prescribed in the Saccharin Order, 1949 as follows:

A saccharin tablet or other sweetening tablet containing saccharin

(i) shall contain not less than 0·18 grains (10·8 mg) and not more than 0·22 grains (13·2 mg) of saccharin or the equivalent weight of soluble saccharin;

(ii) may contain as excipient sodium bicarbonate with or without other suitable substances, the total amount of excipient not to exceed four times the maximum quantity of saccharin;

(iii) shall not contain more than 5 per cent water-insoluble matter nor less bicarbonate than that required to render the saccharin completely soluble.

The latter prohibited the use of artificial sweeteners other than saccharin, in the composition or preparation of any food sold or intended for sale for human consumption. Artificial sweetener was defined as any chemical compound which is sweet to the taste, but does not include saccharin, any sugar or other carbohydrate or polyhydric alcohol.

With the coming to an end of the Defence Regulations in 1954 the Government proceeded to bring on to the statute book the consolidating measure known as the Food and Drugs Act of 1955. Present day legislative control of foods, food additives and contaminants rests on this single food law which applies to England and Wales. Scotland and Northern Ireland have their own Acts and make their own regulations, all three sets of regulations being however almost identical.

Before discussing the more general aspects of food legislation and its enforcement it may be pertinent to consider those regulations in force today, related to sweeteners, as being of immediate interest to the audience at this symposium. The Artificial Sweeteners in Food Regulations 1967 superseded the 1953 Orders and specified in regulations 5 and 10 those artificial sweeteners which could be sold for human consumption and which could be used in food intended for sale for human consumption, *i.e.* saccharin, saccharin calcium, saccharin sodium, cyclamic acid, calcium cyclamate and sodium cyclamate. Regulations 6 and 7 laid down the requirements as to the composition of artificial sweetening tablets and the labelling descriptions. Thus full strength tablets were to contain solely either 11 to 14 mg of saccharin or its salts or 183 to 233 mg of cyclamic acid or its salts. Half-strength tablets were to contain solely 5·5 to 7 mg of saccharin or 92 to 117 mg of cyclamate. Tablets containing a mixture of saccharin and cyclamate had to contain not less than 1 mg saccharin and 15 mg of cyclamate.[4]

Following the banning of cyclamic acid and its salts as permitted food additives, consequential alterations had to be made in the 1967 Regulations. The present Regulations which came into force on 1 January, 1970 are the Artificial Sweeteners in Food Regulations 1969. They are substantially the same as the 1967 Regulations except for the deletion of cyclamic acid and its salts from the list of permitted artificial sweeteners or permitted ingredients in artificial sweetening tablets.[5]

Some 20 Regulations and 19 Orders in our food legislation deal with the composition and labelling of individual food items. Among them only the Soft Drink Regulations of 1964, as amended in 1969 and 1970, are of importance in considerations of the use of permitted artificial sweeteners. The 1964 Regulations allowed in the compositional standards for soft drinks, other than semi-sweet soft drinks for consumption without dilution, the presence of a maximum of 56 gr saccharin and 933 gr cyclamic acid in 10 gallons except in brewed ginger beer and herbal and botanical beverages where the respective quantities were 80 gr and 1333 gr. Soft drinks, other than semi-sweet soft drinks for consumption after dilution, could contain a maximum of 280 gr saccharin and 4666 gr cyclamic acid per 10 gallons.

Semi-sweet soft drinks for consumption without dilution were allowed a maximum of 28 gr saccharin and 466·5 gr cyclamic acid per 10 gallons. The 1969 amendment removed cyclamic acid and its salts from the list of permitted ingredients in soft drinks while the 1970 amendment deleted brominated vegetable oils from the composition of soft drinks.[6,7,8]

THE FUNCTIONS OF THE FOOD AND DRUGS ACT, 1955

This brief survey describes specifically the situation existing in relation to the legislative position of artificial sweeteners in this country and may be regarded as a suitable utilisation of the general principles enshrined in our food legislation. The two basic objectives of the Food and Drugs Act 1955, its 6 Parts, 137 Sections and 12 Schedules are embodied in Section 1 and Section 2. Section 1 protects the consumer against injury to his health by making it an offence to sell for human consumption any food to which substances have been added (or which has been processed in such a way) so as to render it injurious to health. Thus if sugar contains ground glass, or there is arsenic in beer, or cream contains a toxic preservative, that renders it *prima facie* injurious to health. Section 2 protects the consumer against fraud by making it an offence to sell to the prejudice of the purchaser any food not of the nature, substance and quality demanded. If, therefore, there is sawdust in wholemeal flour, added water in milk, or some synthetic substitute for what is ostensibly fresh egg in a particular product, it is not 'as demanded by the purchaser'.

Responsibilities for the implementation of the Act are shared by the Ministry of Agriculture, Fisheries and Food, the Department of Health and Social Security, the Scottish Home and Health Department and the Department of Health and Social Services (Northern Ireland). Under powers conferred on them by Section 4 of the Act,

Ministers may make Regulations which seem to them 'to be necessary or expedient in the interests of public health, or otherwise for the protection of the public' for a number of reasons such as

(*a*) for requiring, prohibiting or regulating the addition of any specified substance to food intended for sale for human consumption,
(*b*) for requiring, prohibiting or regulating the use of any process or treatment in the preparation of any food,
(*c*) for prohibiting or regulating the sale, possession for sale or importation for sale of food not complying with our regulations.

Section 4 also requires Ministers to have regard to the desirability of restricting so far as practicable, the use of substances of no nutritional value as foods or as ingredients of foods. All foods sold in the UK for human consumption are subject to the requirements of this section, with the single exception of milk, for which separate controls are imposed, although cream (and any food containing cream or milk) is covered by Section 4. It is worth mentioning that any food additive regulations made under Section 4 do not restrict the freedom of food manufacturers. They define, in fact, for the trade, those areas of unfettered operation for which the general provisions of Section 1 might otherwise create uncertainties.

Since microbial contamination and spoilage are as much of a risk to health as the presence of intentional and unintentional additives, our Act of 1955 also contains provision in Sections 13 to 25 for promulgating regulations controlling hygiene in the preparation, handling, storage, transportation and serving of foods. Section 8 makes it an offence to sell food unfit for human consumption, so defined as to include microbiological unfitness.

THE NEED FOR FOOD REGULATIONS

The provision of adequate nutrition and safe food for vast populations invariably brings about a proliferation of food regulations. Any mishaps or malpractices are today much more liable to result in disaster of epidemic proportions with heavy calls on local health services. Therefore a clear need exists for some form of control in the interest of public health and for the protection of the consumer. But regulations can only achieve success in the last resort if complemented by social responsibility exercised by the food industry and by increasing public awareness of the proper use of the available food. Often a sense of social responsibility suffices but the needless irresponsibility of a minority frequently justifies the institution of legal restraints.[9]

Failure to provide nutritious food products may result eventually in the development of nutritional deficiencies and consequential injury to the health of either the community as a whole or of certain vulnerable sections, such as the children or the elderly. Consequently there has been an urge on governmental authorities to introduce regulations controlling the composition and labelling of particular classes of foods. These include either staple items of the national diet such as bread, flour, butter, cheese, milk, meat products, etc. or food items consumed particularly by children such as ice cream, soft drinks, preserves, etc. Compositional regulations may incorporate compulsory food additives for the purpose of nutritional supplementation such as the addition of iron, vitamin B_1, and nicotinic acid to white flour or the fortification of margarine with vitamins A and D.

Despite the general purpose of food regulations to ensure that the individual consumer has access only to safe foodstuffs which will not be injurious to his health, they can neither control the foolish nor the abusers of food, whether from pathological abuses or through deliberate action. There are five major areas, however, in which food regulations can play a role in reducing or removing dangers to health arising from foodstuffs and thus contribute to the safety of the consumer:[9] (i) control of food hygiene in relation to microbiological contamination and parasite infestation, (ii) control over unintentional contaminants, (iii) possible elimination of naturally occurring toxins, (iv) control of irradiation of food, and (v) control over intentional food additives.

FOOD HYGIENE CONSIDERATIONS

Today the vast majority of food-borne poisoning is caused by contamination with micro-organisms rather than a chemical. Food scientists with their intimate knowledge of the technology and microbiology of different foods are primarily responsible for the hygienic safety of manufactured products, while our food regulations specifically exercise their preventive functions by establishing machinery for inspection, control and enforcement of general standards of hygiene over various broad areas of food handling and over food categories rather than in relation to individual foodstuffs. Thirteen food hygiene regulations are on the statute book at the moment.

Imported food is subject to similar controls applicable to all articles of food landed in the UK as part of the cargo of a ship, aircraft or hovercraft. It is an offence to import food for sale or human consumption, which is unfit for this purpose or is unsound or unwholesome. Any food in contravention of these requirements may be taken before a justice of the peace who may order its condemnation.

Food regulations are complemented in their preventive functions by some codes of practice which may be regarded as unofficial standards agreed between the trade and the Governmental authorities. They deal with those aspects of food labelling, description and hygiene that for certain reasons are unsuitable for statutory regulations. Although without direct legal authority, these codes of practice are accepted by the courts as evidence of good commercial practice, thus having much of the force of law even when applied to imported products. Some seven codes of practice are in existence at present.

No specific legislation exists in the UK in the field of parasite infestation of food although parasites are still the source of considerable human morbidity. Even less is known about the threat to human health from protozoal parasites such as toxoplasma. The great technological difficulties in detecting carrier foods and the practical impossibility of controlling eating habits of the population at large vitiate any attempts at devising sensible legislative controls. Moreover, the hazards from animal parasites to human health are less generalised than those arising from micro-organisms because of the former being related more closely to traditional eating habits and local or national dietary preferences. The only other avenue for useful Governmental activity in relation to parasites and microbial infestation of food presents itself in the educational field. This aim cannot be pursued by the framing of compulsory legislation but is nevertheless an important task of governmental organisations concerned with public health. It involves the making available, by the usual channels of communication, of all useful information that would explain in simple, easily understood terms the important issues involved, such as the necessity for proper refrigeration and cooking and the dangers of cross-contamination, to all those engaged in the handling and preparation of foodstuffs, without frightening the consumer about specific food products.[10]

It can be seen from the foregoing summary that food legislation does not, in its attempt to ensure safety in the field of food hygiene, cover all factors that may cause food to be a hazard to health. Existing standards deal only with micro-organisms responsible for infectious disease or for the formation of toxic compounds in food but they do not extend to viruses, animal parasites and fungal toxins (except for aflatoxin in some foods).

UNINTENTIONAL CONTAMINANTS

Many classes of potentially toxic agents which could enter food occur in the environment to which the general population is exposed. Pesticides, drugs, cosmetics, household chemicals, industrial chemicals, radioisotopes, natural chemicals of non-food origin, atmospheric and water pollutants, are all potential or actual hazards to man in modern sophisticated society. Much of the food hygiene legislation is designed to reduce the chances of food poisoning happening from such sources but the occasional slip achieves its due notoriety. It is clearly impossible to legislate for every possibility.

The situation differs somewhat for pesticides, processing residues, packaging migrants or environmental radioisotopes. Insofar as pesticides protect foodstuffs from damage and spoilage they have become indispensable to modern agriculture. In consequence of their application to crops during growth and post-harvest storage, residues of these substances tend to persist in the final food but not in all cases. As pesticides are designed to be inimical to certain forms of life they are intrinsically toxic to a variable degree and the toxicity of residues needs to be assessed. Other contaminants arise from modern developments in food technology resulting in the presence of small amounts of unintentional migrants in food offered for sale. Yet others, such as solvent residues, may be present in food as a result of ancillary processes necessary in the preparation of food.

Much controversy exists over the best way to regulate in food for sale the size of these residues, usually minute amounts in themselves. Direct legislative control by a multitude of specified 'statutory' tolerances is practised in many countries. We in the UK have been notoriously unwilling to join the league of adherents to statutory tolerances because the total diet studies undertaken over recent years by the laboratory of the Government Chemist have shown, for instance in relation to the persistent organochlorine compounds, no progressive increase in the levels of some and a fall in others. In any event, if foodstuffs are found with pesticide residues exceeding those deemed desirable, it is a short-sighted policy to order confiscation. The urge to introduce statutory tolerances for certain pesticides may be a reassuring political move but has medically little to commend itself. It has been our experience, also borne out by the results of extensive and continuing monitoring of food for pesticide residues carried out nationally, that there is negligible danger from this source provided good agricultural practice as recommended for each pesticide is carried out. Furthermore we believe that the control of pesticide usage is a more effective safeguard of the health of the public than the enforcement of arbitrary tolerances set in good faith on evidence that is often difficult to evaluate in terms of risk to human health.

Only a selected few environmental contaminants, *i.e.* arsenic, fluoride and lead, are presently being controlled by our food regulations. The control over the use of antibiotics in animal feeding-stuffs indirectly restrains the residues likely to be found in food and the recent recommendation of the Swann Committee[11] to allow only the use of feed antibiotics which have no therapeutic role and value, further safeguards the consumer against possible adverse effects as a consequence of the use of antibiotics in medical therapeutics. The strict control and surveillance of radioactive materials production and their disposal ensures that the radioisotope content of our food does not exceed significantly the unavoidable contamination from atmospheric pollution.

Contaminants can only be controlled successfully if they are clearly identifiable and can be estimated accurately. Usually governmental control is exercised by the establishment of permitted (positive) lists implying that anything not entered in such a list is automatically excluded. Lists of forbidden contaminants normally suffer from the drawback that everything not included in the official definitions of the listed substances is, by implication, permitted. Such a negative list gives no guarantee that other noxious substances, not known or identified at the time of its promulgation, have not been overlooked and are able to continue to exert their deleterious influences on the health of the public unchecked. In order to be included in a permitted list the safety of the compound is scrutinised along the lines which will be described in detail in relation to deliberate food additives. However, it is impossible to test exhaustively every compound likely to become an environmental contaminant of food, the size of such an operation being beyond the capacity of all available testing resources. Even if this were not so, the diversion of scientific manpower and laboratory facilities to the performance of such tests would be unjustified in relation to the smallness of the risk to health of the public.[9] Therefore some compromise is usually made between the desire to base safety evaluations on satisfactory and interpretable experimental data and the practicability of obtaining all the required information.

NATURALLY OCCURRING TOXINS

A widespread but erroneous belief exists that naturally occurring substances are necessarily safer than synthetic substances. Many powerful toxic agents occur as components of natural foodstuffs in plants, animals, and micro-organisms, *e.g.* oestrogens, radioisotopes, trace elements, lactones, coumarins, mycotoxins, plant hepatotoxins, etc. Most of these natural substances possess very complex chemical structures not easily dealt with by the normal metabolic processes of animals and man. Gaps exist even in our knowledge of the constituents of human food and their variations with environment or genetic make-up of animals and plants. The list of toxic substances found in nature continues to grow rapidly and the occurrence of traces of these in human food is likely to become a greater health problem than the use of deliberate additives. The latter, at least, are investigated toxicologically but the natural toxins are not, and it is unscientific to argue that substances that have been consumed over centuries must be safe.

IRRADIATION OF FOOD

In most countries, including Britain, there exists machinery dealing with the application of ionising radiation to food intended for human consumption, based on exemption to be sought for specific processes and applications from total prohibition. In Britain the safety evaluation is based on a memorandum issued by the Advisory Committee on the Irradiation of Food[12] setting forth a schedule of the information to be submitted. It includes a full description of the food, its packaging, the proposed method of irradiation, the safety precautions, storage, tests for residual radioactivity, nutritional quality, the toxicity of irradiated food including carcinogenicity and mutagenicity, microbiological flora before and after treatment and detectability of treatment.

INTENTIONAL FOOD ADDITIVES

Many of our numerous food regulations deal directly with additives in food and their safe use. Some seven technological classes of additives have already been covered by specific regulations while flavouring agents, packaging materials, acids, bases, humectants, propellants, release agents, etc. represent significant gaps in our attempts at creating a regulatory fence although they remain subject to compliance with the general provisions of Sections 1 and 2 of the Food and Drugs Act of 1955, and therefore do not escape statutory control entirely. There has been, however, a considerable increase in recent years in the number and nature of available food additives. Modern food processors, in response to pressures of present day demands, have come to rely increasingly on chemical additives. Despite the often emotional criticism, these additives are not simply the mischievous devices of food manufacturers intent on deceiving the consumer. They confer real advantages by reducing food wastage, by allowing wider distribution and better efficiency of manufacture. Thus it is more acceptable to most consumers to have their margarine containing an antioxidant than to have it rancid. Yet not all additives have the same command of virtue. Whether the addition of a nutritionally inert cellulose ether to, say, bottled mayonnaise as an emulsifier, makes it as acceptable as the natural ingredients is open to more doubt, and there are some who would vigorously question whether artificial colourings have any justification at all—except to provide three-tone printing experts with opportunities to display the results of their arts and crafts in the women's glossy magazines. In short, all deliberate food additives are not of equivalent righteousness, though many are technologically of major importance in food production.

Food additives may be defined as substances, whether naturally derived or deliberately synthesised, which are added intentionally to food materials in order to effect some change in their properties to the advantage of the consumer without being foodstuffs in themselves. Now food, like all other matter, consists of nothing but chemicals. Many food additives already occur in food, others are usually not part of our diet. As the basic technology merely involves the addition of a small amount of one chemical to a much larger mixture of many different chemicals, there is nothing inherently alarming or unpleasant in this procedure,[13] provided those principles governing the use of intentional food additives are adopted, which have been elaborated by the first meeting of the Joint FAO/WHO Expert Committee on Food Additives.[14] These principles are:

(*a*) a food additive should be technologically effective,
(*b*) a food additive should be safe in use,
(*c*) a food additive should not be used in any greater quantity than is necessary to achieve the technological effect,
(*d*) a food additive should never be used with the intention of misleading the consumer as to the nature and quality of food,
(*e*) the use of non-nutrient food additives should be kept to the practicable minimum.

In determining the eventual form adopted in framing food additive regulations the above principles are taken into account as well as the fact that food additives may be taken by large sections of the population over prolonged periods of time and that, unlike drugs, their use is not under direct medical or other personal supervision. Another consideration is the existence of two valid but opposing views on the use of intentional food additives. One school of thought accepts only the minimum number of additives consistent with technological needs but resulting, as a consequence, in larger individual intakes of each permitted chemical. It claims in addition a reduced probability of biological interaction between additives themselves, between additives and components of food and between drugs and additives; all useful because of the practical impossibility of predicting these hazards, thus virtually precluding their control. The other school of thought believes in accepting a large number of additives with consequent lower intake of any individual additive.

SAFETY TESTING OF FOOD ADDITIVES

Deliberate food additives are not intrinsically toxic. Enormous quantities of single doses would have to be ingested to produce adverse effects but uncertainty prevails about the ultimate effects on man of ingesting small amounts daily over many years. Such insidious possibilities are not easily checked by direct human observation because human experiments are possible only on a very small scale and do not mimic the usual life-long exposure to very low doses. Even epidemiological surveys would be 'locking the stable door after the horse has bolted'. Accordingly, tests are carried out in animals, usually rats and mice, sometimes dogs, and occasionally other species. They are fed at levels far in excess of those to be used in human food, both for short periods and also throughout their lifespan. Any changes in a large number of physiological parameters are noted and tumour incidence is recorded. In extrapolating the results to man it is usually accepted that no harm arises if a substance cause no detectable effects at a level of at least one hundred times the maximum amount likely to be added to human food. This sort of information can never give an absolute assurance of safety for a substance unless supplemented by careful study of the effects, or lack of them, of ingestion by persons of all ages over long periods.

There is a surprising amount of international agreement on the basic information required by governmental authorities concerned with the health of the public. Many valuable publications have appeared which discuss in detail the minutiae of the safety investigations of food additives. To allow useful comparison and evaluation of biological studies on food additives it is necessary to have proper identification and a satisfactory specification of the substances being tested. Toxicological studies are valueless in assessment if there is no guarantee that the substance tested is the same as that used in the food.

It would be inappropriate in the context of my talk to discuss the various types of toxicological studies undertaken for safety evaluations because it is not one of the functions of food regulations to lay down exact schemes to be followed in toxicological investigations. This is entirely the province of the investigator, as any attempts at imposing rigid schemes of testing by regulations lead to waste of valuable laboratory resources and scientific manpower. Nor should the performance of toxicological investigations be regarded as a do-it-yourself operation. Its very relationship to human health necessitates the overall guidance by scientists possessing the required expertise.[9]

Although the types of animal and eventual human investigations will vary according to the nature of the test compounds, certain standard experiments such as LD50, 90-day tests and lifespan studies are almost always carried out in order to gain some conception of its toxic potentialities. As the number of possible investigations is very large and the choice of species often bewilderingly varied, it is vital to be selective in order to be able to complete the assessment within a reasonable time and at an economic cost. Sufficient well-recognised biological parameters require study to permit the determination of levels at which food additives are safe in several animal species, to gain some idea of the metabolic fate and to detect any carcinogenic, teratogenic and mutagenic potentials.

EVALUATION OF RESULTS

After completion of the experimental programme there remains the vital task of transferring the observed results to man. Normally an estimate of safety is arrived at by finding the maximum no-effect level in the various animal species examined. If the material has been shown to have carcinogenic properties, any such substance is *ipso facto* completely unsuitable for use as a food additive. However, the significance for man of some tumour findings in animals may be difficult to assess, especially if these appear following administration by routes other than an oral one.

When a no-effect level has been established it is necessary to provide an adequate margin of safety between the highest no-effect level and the amount to be permitted in food. Such an estimate has to take into account a number of factors. Man may be more sensitive than the species of animals used for testing. Humans like any other species vary in their sensitivity to foreign substances. The young, the old, the sick and the healthy may respond in different ways to the same chemical stimulus. Infants often lack the enzymic complements for dealing with substances which present no problem to the healthy adult. Genetic defects may be a handicap for small sections of a population and pathological alterations in metabolic or excretory capacity may lead to accumulation of foreign materials. It must be realised that any substance may produce effects on the body if consumed in sufficient quantities or under particular circumstances, *e.g.* monosodium glutamate causing the so-called 'Chinese Restaurant Syndrome'.[22]

The biological effects of compounds vary with the dosage administered and this relationship is fundamental to safety evaluation. To allow for this and for the other possibilities previously enumerated, the no-effect level determined in animals is regarded as a hundred times greater than the level to which man should be exposed. If the available toxicological data are deficient, safety factors of 1000 or even 2000 may be used. On the other hand, where extensive investigations on man are available the factor may be reduced to 50 or lower.

Therefore the acceptable daily intake (ADI) for man is arbitrarily set at one-hundredth of the maximum no-effect level determined in the most sensitive mammalian lifespan study. The precise choice of the safety factor will depend on such other considerations as the types of food in which the food additive is used, whether these are staple foods, or with big seasonal variations in consumption, or whether they are consumed by children or other specially vulnerable sections of the community. The ADI is thus simply a biological measure of the maximum amount of a substance that could be consumed throughout life without demonstrable ill effects and no great hazard occurs if the ADI is occasionally exceeded.

The concept of an ADI is of great importance for the task of regulatory authorities engaged in examining the actual dietary intakes of food additives. The aim is to exercise appropriate control by either limitation of use of a food additive or regulation of composition of foodstuffs with adequate measures for surveillance. It is thus easy to understand why food additive legislation takes the form of permitted lists of additives included on the basis of safety evaluation and adequate evidence of technological need.

Complications have and still do arise scientifically and administratively because many food additives were in use before safety testing was applied as a condition of approval for use. It is then awkward to recommend that a particular compound which has been employed for many years in this way without obvious hazard should be banned until it has been subjected to the whole gamut of animal testing. The compromise which tends to be adopted is to condone a period of grace, during which time the animal tests are to be completed. But it also provides a favourite target for the opponents of food additives.

ADVISORY COMMITTEES

Advice about the acceptability of the various non-nutritious food additives and generally about matters related to the Food and Drugs Act 1955 is available to Ministers through two bodies of independent experts. One is the Food Standards Committee which deals with compositional standards, and regulations relating to the description, labelling and advertising of food. The other is the Food Additives and Contaminants Committee (FACC). Both are serviced by the Ministry of Agriculture, Fisheries and Food. Membership is drawn from industry, universities and from Departments all serving in a personal capacity and not as representative of any particular interests. Both Committees may refer questions of safety-in-use of foods and additives to the Pharmacology Sub-committee of the Committee on the Medical Aspect of Food Policy serviced by the Department of Health and Social Security. These latter bodies are similarly composed of independent academic and medical experts and departmental assessors.

With contaminants, the aim of the FACC and Government departments is to encourage good manufacturing practice to reduce contamination of food to the lowest possible level. With additives, the Committee has to be satisfied that there is a distinct technological need or benefit to the consumer, that the proposed use of the additive does not entail a hazard to health and that satisfactory specifications for food use exist. Advice may be sought from many quarters, *e.g.* research associations, industrial experts or representatives from the industry directly involved. After a conclusion has been reached a report is sent, usually to the Minister, including a report from the Pharmacology Sub-committee on the biological evidence submitted for evaluation. In this report recommendations are made for each additive requested in the industrial submissions, a specification is given for each approved substance and, where appropriate, limits are set on the amount that may be added to food. In reaching decisions Ministers act in the main on the proposals of these Committees, after inviting comments from the industries concerned. However, Ministers are not obliged to wait for the advice of these committees and they can accept or reject it as they see fit.[23]

The general policy with food additives is to produce statutory permitted lists for all classes of additives and, possibly, contaminants, and thus to ensure that eventually no additive which is not on one of the permitted lists may be added to food. However, it is not intended that these permitted lists should be handed down like the proverbial ten commandments but it is proposed to review each class of additives approximately every five years. Once Ministers have agreed to such a review, comments and evidence are invited from all interested parties. The Committees themselves have no facilities for testing and therefore rely on the reports of work carried out elsewhere in competent laboratories.[23]

The use of many additives dates from a period when no requirements existed for routine toxicological testing. Thus some of the traditional additives now find themselves under notice to be investigated for evidence of safety-in-use before the next survey of their class. Even long established practices must be critically reassessed in the light of new advances in biological testing. On the other hand, an early decision on the need for a proposed new additive may save a good deal of money otherwise spent on unnecessary testing and avoid wasting precious facilities if the grounds for need are considered inadequate.

MAKING REGULATIONS AND THEIR ENFORCEMENT

The safety assessment of a food additive is only valid within the limits of the accuracy and completeness of the scientific evidence upon which it is based.[24] If the substance tested differs significantly from the specified material actually used in food then the evaluation is valueless. New knowledge or better methods of analysis may alter the interpretation of previous findings, *e.g.* the discovery of an unsuspected metabolite. Such changes in assessment may raise serious problems for industry, distribution, governmental authorities or for the consumer. Legislative machinery must exist to permit implementation, with reasonable speed, of any changes in safety evaluation as a result of new discoveries. Such action may be taken in the UK by means of Amending Regulations which become legally effective the day they are laid before Parliament but which are subject to negative resolution and may be prayed against, in which case there is a debate and possibly modification during the next 40 days of parliamentary business.

A few words remain to be said about the judgement of governmental authorities on what and when to test. Related to this activity is the concept of 'levels of toxicological insignificance'. For every substance there is some low level when its ingestion is safe or when the possibility of hazard becomes so remote that it is superfluous to institute government regulations to protect the public. A recent report to the Food Protection Committee of the National Academy of Science of the USA[25] developed some useful guide lines for arriving at an estimate of a toxicologically insignificant level for a food additive or contaminant and made the not unreasonable suggestion that at levels of 0·1 ppm in the diet, the majority of chemicals, with the exception of some heavy metals, pesticides, drugs, carcinogens and other biologically active organic compounds, represented no toxicological risk to the consumer and could therefore be ignored from the regulatory point of view. Once the Advisory Committees have sent their reports to Ministers their task is normally completed. The responsibility for the final decision, and for drawing up regulations, rests entirely with the Ministers concerned, even though the report may have been published and comments have been invited. In the light of representations received the relevant Government Department prepares proposals for regulations which are again subjected to a round of comments before final regulations are drafted and laid before Parliament. When accepted by Parliament regulations usually come into force after a time interval sufficient to allow the trade to carry out the necessary changes in formulations and labelling. Despite the apparent laboriousness of the procedure, it offers ample opportunity for interested parties to make representations and for full consideration of all points of view.[23]

Enforcement of the main provisions of the Food and Drugs Act 1955 and the regulations made thereunder are the statutory duty of the Food and Drugs Authorities. These are all the local authorities in the UK and the Port Health Authorities at airports and seaports, but there is no central law enforcement authority as exists in the American FDA or Canadian FDD.

CONCLUSIONS

Reviewing the present situation leads to the conclusion that too much or too little food is more harmful than all the additives, intentional or unintentional, likely to be found in food. It is also obvious that intelligent appraisal of all available toxicological and technological data is the only way to arrive at a scientifically valid decision on the safety-in-use of a food additive. There is very good evidence that infections conveyed to man by foodstuffs are a greater hazard than all food additives when properly used. Thus the eventual estimate of the risk arising from the use of a specific additive should form the real basis for action by governmental authorities, whether in the form of legislation or official recommendations. On the whole our food regulations, despite all their imperfections, are effective in safeguarding the health of the population against hazards from additives as long as there exists also a concomitant attitude of social responsibility on the part of the food manufacturing and food distributing industry. However, regulations without controls or sanctions are worse than no regulations at all.

REFERENCES

1. Coomes, T. J. (1970). In *Conference on the Layman's Guide to the Food Regulations, RSH* (in press).
2. *Stat. Instrum.*, 1953 No. 1310.
3. *Stat. Instrum.*, 1953 No. 1311.
4. *Stat. Instrum.*, 1967 No. 1119.
5. *Stat. Instrum.*, 1969 No. 1817.
6. *Stat. Instrum.*, 1964 No. 760.
7. *Stat. Instrum.*, 1969 No. 1818.
8. *Stat. Instrum.*, 1970 No. 1597.
9. Crampton, R. F. and Elias, P. S. (1970). In *Conference on the Layman's Guide to the Food Regulations, RSH* (in press).
10. WHO (1968). *Wld. Hlth. Org. Techn. Rep. Ser. No.* 399.
11. *Swann Committee Report* (1969). Joint Committee on the use of Antibiotics in Animal Husbandry and Veterinary Medicine. HMSO, London (CMND), Cmnd 4190.
12. *Ministry of Health* (1968). Report of the Advisory Committee on the Irradiation of Food, HMSO, London.
13. Frazer, A. C. (1968). *Brit. Nutr. Found. Inf. Bull.*, **2,** p. 32.
14. WHO (1957). *Wld. Hlth. Org. Techn. Rep. Ser. No.* 129.
15. WHO (1967). *Wld. Hlth. Org. Techn. Rep. Ser. No.* 348.
16. WHO (1958). *Wld. Hlth. Org. Techn. Rep. Ser. No.* 144.
17. WHO (1961). *Wld. Hlth. Org. Techn. Rep. Ser. No.* 220.
18. Philp, J. McL. (1968). In *Modern Trends in Toxicology*, Ed. by Gouldin, R. and Boyland, E. Butterworths, London, p. 243.
19. MAFF (1965). *Memorandum on Procedure for Submission on Food Additives and on Methods of Toxicity Testing*, HMSO, London.
20. DHSS (1968). *Report of the Consultative Panel on Carcinogenesis*, HMSO, London.

21. Elias, P. S. (1968). *Congresso sullo tutela sanitaria degli alimenti Firenze*, 2–5 May, 1968.
22. BIBRA (1968). *Inf. Bull.*, **7**, p. 376.
23. Weedon, B. C. L. (1970). *Chem. in Britain*, **6**, p. 242.
24. Frazer, A. C. (1968). *Brit. Nutr. Found. Inf. Bull.*, **2**, p. 21.
25. NAS/NRS (1969). *Report of the Food Protection Committee, Food & Nutrition Board, National Research Council, Washington.*

DISCUSSION

Harper: As someone with a sweet tooth, waging an ever-losing battle against excess weight, I viewed with considerable personal regret the decision to remove cyclamates from the market, which forced me to change over with reluctance to the more bitter-tasting saccharin. Now I see in one of the recent issues of FDC reports (12.4.71) that there has been an announcement at the American Cancer Society in Phoenix by Dr George Bryan, to the effect that he had repeated his findings that saccharin, implanted in the bladders of mice, was carcinogenic. This claim has been supported by one of the chief scientists at the National Institute, Dr Umberto Soffioti, so it looks as though I may soon have to face up to the choice of either contracting bladder cancer or dying of coronary heart disease. At the risk of repeating points that may have already been made by former speakers, the FDA really had no choice, when faced with the Delaney Clause, but to ban cyclamates following the carcinogenicity finding. It has been argued that they could equally have enforced the ban, and with greater legal validity, by invoking the main Food Additives Law which permits only items that are proven safe to be allowed in the food supply. This would then have placed the onus on the manufacturers to provide evidence of safety and refute the carcinogenicity claim. However, it was the United States ban and the adverse publicity from uninformed sources that led to other governments taking the same decision. I happen to believe it was the correct one, for otherwise, what is the purpose of performing toxicological tests on animals, if we are then going to disregard only the unpalatable results and suggest that animal experiments really have little significance in relation to the situation that exists in regard to man.

Our own laboratory has conducted very extensive studies into the toxicological and metabolic effects of cyclamates and in no case have we recorded what we considered to be a significant user hazard. In the face of the mass of similar data attesting to its safety, I also believe that the FDA were right not to take action on the strength of the chromosomal damage reported by Legator, and the chick embryo malformations reported so sensationally by Jacqueline Verrett, both of the FDA. In neither case do we have any real evidence attesting to the human significance of such tests, but I would submit that an entirely different situation arose when the confirmed induction of bladder tumours in rats came to light. To my knowledge, most, if not all of the known human bladder carcinogens have been found to be capable of the same effect in one or more animal species, so that this finding could not be ignored. They might, I suppose, have delayed action pending the outcome of repetitive or additional studies, but with about 1 billion dollars worth of food containing cyclamate being consumed in the United States and increasing all the time, I assume that they did not consider the risk to be justified, a view with which I concur. It remains for future work to determine whether or not the present ban can be relaxed or even removed.

George: One thing Dr Elias did not mention is the Memorandum on Procedure for Submissions on Food Additives and on Methods of Toxicity Testing (HMSO 1965) which is a guide to those who wish to submit a food additive for inclusion in the Regulations.

Green: 1. Why must there be separate regulations for Scotland and Ireland?

2. Many foods are approaching a therapeutic character. How will they be defined into one or another group?

Elias: 1. This is probably political.

2. The definition of what is a therapeutic substance? I cannot give it to you, but it is now written into the Medicine Act of 1970. I think the basic point is that it has got to have some therapeutic action, *i.e.* nutrition alone is not good enough.

Open Forum

Stacey: I do not believe the story of dental caries with respect to sucrose. Surely one would get far more polysaccharides in glucose syrups and thus get more microbial growth. Will not the problem become greater by changing from sucrose to glucose syrups?
Selby: The evidence is against this. The experience in fermentation of sucrose is that it elaborates odd polysaccharides which are difficult to control. Experimental dental evidence supports the advantages of glucose syrups.
?: Glucose and glucose syrups, though not as cariogenic as sucrose, are, nevertheless, not without cariogenic properties. This whole problem of selling sweetness to the public, seems to be one associated with the dental health of the population set against economics.
Grenby: Sucrose produces more tooth decay than other sugars. Oral micro-organisms produce acid from dextrose, but both acid and polysaccharides (dextrans) are produced from sucrose. Dextrans help to form dental plaque on the teeth, which holds acid right up against the tooth enamels initiating decay.
Spence: It is my belief that the use of dextranase may enable us to consume sugar and still dispose of the problem of dental plaque formation, the precursor of dental caries. This would nullify Dr Selby's contention of one advantage for his wheat sugars.
Stacey: Is synergism a real thing? If so, honey should be the sweetest thing on earth since it contains 26 identifiable sugars.
Birch: Cameron has given evidence in support of synergism. A large number of sugars does not necessarily lead to a large synergistic effect.
Johnson: There are parts of the world where honey is toxic. Its composition is not constant. It contains essential oils from flowers which will alter the taste of the sugars, and polysaccharides which may depress the sweetness.
Stacey: It is, indeed, a complex sugar mixture.
Birch: Dr Siddiqui reviewed the subject recently for 'Advances in Carbohydrate Chemistry'.
Clarke: Will Dr Elias enlarge on the toxicologically insignificant level of

0·1 ppm in the diet? 2·5 million tons of sucrose constitute 30% of the national diet. Does this mean that an additive used at less than 0·75 million tons per annum can be regarded as toxicologically safe?
Elias: It is not dependent on the tonnage used. The concentration in the diet is the criterion, for instance, the level in the United States is 0·1 ppm.

In terms of low ppm levels in the diet, these amounts may be similar to those normally encountered in one's environment which contains potent carcinogens, *e.g.* lead in air. It is no good to say foods should be free from such as they never will be, due to the environment. The determination of the toxicologically insignificant level is difficult due to the cost of testing, especially if the level of use is small. If the chemical structure is inoffensive and the proposed level of use is only 0·1 ppm, then complicated studies are not called for, but we do need to know areas of use and also who will consume them. For instance, if children will be the consumers (soft drinks for example), a more rigorous examination is needed.

Food packaging materials are so complex that it becomes too involved to apply the full gamut of tests.
Crampton: This question has been debated for the last three years in both the United States and the United Kingdom, by the FDA and members of committees advising the ministry. It is unlikely that any more data will become available within the next year or so to enable a decision to be made. Industry must be able to plan ahead and has to perform toxicological tests. How and when will it be possible to give an answer to the question? Have levels of toxicological insignificance been approved and, if so, what are they?
Elias: It is difficult for a Government Department to come out with a statement of this nature. If industry gives details of what compound will be used, and in which foods, then some latitude may be allowed by giving an informal classification. It is impossible to make a general statement on non-toxic levels, or to define safe levels of pesticides, carcinogens, etc. No committee can say before testing has been done.
Gourley: Would Dr Elias please comment on the effectiveness of control of accepted daily intake (ADI) for additives used in a number of different foods and also how allowance is made for introduction of new products. What happens for instance, when the ADI has been set, and a new application comes along which raises the level of usage?
Elias: The committee is not enamoured of the ADI concept. It is regionalised even within a country, therefore such calculations must be based on very gross assumptions. The committee prefer to have recommended levels rather than ADI's. For instance, in wine drinking countries, the whole of the ADI for SO_2 is used up by the wine leaving nothing for other foods.
Summers: I gave up sugar in tea years ago, now I find it distasteful. Is there a psychological explanation for this?
Watson: The pleasure of a given food or drink is partly a function of the situational and emotional factors operating at the time. After an initial 'adverse' period, the unsweetened tea now achieves pleasurable status from the accompanying pleasurable settings. Should sucrose be added

subsequently, the stimulus pattern no longer complies with that which has been associated with pleasure—hence it is unpleasant.

Spencer: What was the sample size used to produce the conclusion that extroverts have higher sugar preferences?

Watson: 30–50; extroverts like a general increase of stimulus. This does not apply equally to women due to a cultural overlay such as slimming.

Spencer: Is there a parallel with smoking?

Watson: Not the same; sugar and smoking have aspects enhancing social interaction which is what extroverts require.

Stacey: Can Mr Spencer tell us whether taste panel judges are allowed to eat before tasting?

Spencer: They do not taste sooner than 1 hour after a meal.

Stacey: Why is it that we eat the dessert as a last course in a meal, whereas children demand their dessert first?

Spencer: In France, cheese and dessert come in the reverse order.

Watson: I think this is a puritanical aspect of the English character: something that does you good cannot be pleasurable, therefore one eats the beneficial part of the meal, such as protein and vegetable, first and leaves till later the directly pleasurable foods, the preparation of which gives scope for making an elaborate fancy appearance. It is also a means of not reducing the consumption of the more expensive meat dishes.

Crampton: Surely there are more mundane reasons; when there used only to be bread, butter, jam and cream cakes, with the cakes appearing only once a week, then one simply ate the luxury last of all.

Watson: Hard economic times do not dictate the English pattern since continental practices are different.

HRC: It was suggested that, in wine drinking countries, cheese and biscuits are eaten immediately following the main course to use any red wine remaining from the meat course. The white wine then follows with the dessert.

Coulson: 1. Professor Horowitz mentioned that only ozone was successful in removing neohesperidose intact. Have any transglycosylases been examined in this respect?

2. Dr Inglett referred to continental literature on the adverse side-effects of some of the terpenoidal compounds. Some years ago when I was working on lucerne triterpenoidal saponins, Dr Stack-Dunne of the MRC London, found that these saponins and the commercial Quillaja saponin produced potassium retention and anticarbonic anhydrase effects.

Horowitz: The enzyme removes rutinose intact, but leaves neohesperidose, that is the reason why we use ozone. Naringinase, used for debittering, also removes sugars piece by piece.

Coulson: Does Sir Edmund Hirst or the Royal Holloway Group have any evidence of this?

Hirst: No.

Inglett: Both stevioside and glycyrrhizin have steroid nuclear structures and are reported as having toxic action on adrenal function (possibly due to potassium retention). There are undesirable side-effects, though no problem has been encountered with small quantities. FDA forbid use of

large quantities. Stevioside is used for the treatment of Addison's disease.
Elias: On the question of caries and sucrose it is possible that a simple additive such as calcium sucrose phosphate will reduce the incidence of caries without seeking to replace sucrose by some other carbohydrate sweetener.
Stacey: Does it work?
Elias: Trials in sweets have been undertaken along these lines. The confectionery retained its sweetness but its cariogenicity was reduced.
Hirst: It has been suggested that the effect of calcium sucrose phosphate in diminishing the incidence of dental caries, following the use of sucrose in foodstuffs, may be due to the suppression of enzyme activity, which would lead to the formation of polysaccharides in the dental plaque. Extensive experiments have been carried out in Australia which show favourable results.
Gourley: Calcium sucrose phosphate is present naturally in sugar cane and may account for the lower cariogenicity of sugar cane compared with refined sugar, but it is largely removed during the refining of sucrose. ICI are working on this.
Spence: Could I persuade any contributor to say something about the association between the visual quality of foodstuff and the ability to evaluate its sweetness?
Pangborn: The colour of food influences its acceptability and the ability of panel judges to identify its flavour constituents. It is more difficult to do this when the food, *e.g.* jams, jellies, ice-creams, is miscoloured rather than lacking any colour. Several studies have been carried out where miscoloured sweets were tasted and the judges could not identify the flavours. It was also said that red sweets tasted sweeter, the deeper the orange colour the more intense the orange flavour, and the green colour gives the impression of more tartness, due to the connection of this colour with unripeness.
Spencer: This can cause mistakes in panels where appearances may affect the criterion of choice. Due to these effects, it is necessary to standardise the conditions of testing, such as performing under coloured lighting to mask differences.
Pangborn: Relative to colour-flavour associations, the consumer expects foods to have the appearance to which they have become accustomed and develops a resistance to faded or miscoloured products.

Schultz, H. G. (1954) [*Colour in Foods, a Symposium*, Natl. Acad. Sci., Natl. Res. Council, Washington, D.C., p. 186] reported that flavour scores for low-quality orange juice could be improved by adding food colouring. Hall, R. L. (1958) [*Flavour Research and Food Acceptance* (Arthur D. Little, Inc.), Reinhold, New York, p. 391] observed that judges successfully identifying flavours of ice-cream sherbets in their customary colour, had great difficulty recognising white or deceptively-coloured samples. Pangborn, R. M., Berg, H. W. and Hansen, B. (1963) [*Am. J. Psychol*, **76,** p. 492] found that rosé-coloured dry white table wine was considered sweeter than the same wine artificially coloured to resemble sherry, claret, sauterne and burgundy, but only by experienced wine judges, not by naïve panels.

Watson: How much do people think of food as an aesthetic experience rather than a purely nutritious necessity? Why cannot we have gaily coloured foods? In fact, we are working on these problems.

Clarke: Colours are the signposts to the anticipated organoleptic experience and miscolouring can upset people's expectations which leads to incorrect judgements, and often to revulsion and a refusal to repeat the experience.

Coulson: I can vouch for this having met a certain liqueur which was coloured similar to methylated spirits, with the result that I was unable to drink it!

Spencer: A similar experience might explain the British distaste of American root-beers due to methyl salicylate.

Pangborn: The best way to reduce the flow of saliva is to cut off the vision. When a sighted person is blindfolded, he lapses into partial consciousness and the senses become very dulled.

Concluding Remarks

M. Stacey

It is difficult quickly to apply scientific methods to sweetness problems. There will, no doubt, be a time when, due to knowledge of what sweetness is, we shall be able to synthesise the ideal sugar, and although we shall not get the answer quickly, I think Dr Birch is working on the right lines. I am sure that a quantitative estimate of taste-cell reactions will come. Perhaps rather than using sucrose as a sweetener at its artificially adjusted cost, we should use it as a cheap source of organic chemicals!

A great deal has emerged from this symposium. We have had present many representatives from institutes, universities and industry concerned with this problem and this has made for a lively exchange of views. Our thanks must go to the organising committee of the symposium, and to the National College of Food Technology for their hospitality. This has been a particularly pleasurable occasion because we have been able to welcome our overseas visitors, and these distinguished guests have helped to make the symposium a notable success.

Index

C

F

G

N

O

P

Q

R

S

T

U

V

W

Baby - sweeteners - p 23
cyclamate lesioable p 63